ANCIENT EGYPTIAN MASONRY

The Building Craft

By
Somers Clarke and
R. Engelback

Published 1999
The Book Tree
Escondido, CA

Originally Published in 1930

Ancient Egyptian Masonry
ISBN 1-58509-059-X
© 1999
THE BOOK TREE
All Rights Reserved

Published by
The Book Tree
Post Office Box 724
Escondido, CA 92033

We provide controversial and educational products to help awaken the public to new ideas and information that would not be available otherwise. We carry over 1100 Books, Booklets, Audio, Video, and other products on Alchemy, Alternative Medicine, Ancient America, Ancient Astronauts, Ancient Civilizations, Ancient Mysteries, Ancient Religion and Worship, Angels, Anthropology, Anti-Gravity, Archaeology, Area 51, Assyria, Astrology, Atlantis, Babylonia, Townsend Brown, Christianity, Cold Fusion, Colloidal Silver, Comparative Religions, Crop Circles, The Dead Sea Scrolls, Early History, Electromagnetics, Electro-Gravity, Egypt, Electromagnetic Smog, Michael Faraday, Fatima, Fluoride, Free Energy, Freemasonry, Global Manipulation, The Gnostics, God, Gravity, The Great Pyramid, Gyroscopic Anti-Gravity, Healing Electromagnetics, Health Issues, Hinduism, Human Origins, Jehovah, Jesus, Jordan Maxwell, John Keely, Lemuria, Lost Cities, Lost Continents, Magic, Masonry, Mercury Poisoning, Metaphysics, Mythology, Occultism, Paganism, Pesticide Pollution, Personal Growth, The Philadelphia Experiement, Philosophy, Powerlines, Prophecy, Psychic Research, Pyramids, Rare Books, Religion, Religious Controversy, Roswell, Walter Russell, Scalar Waves, SDI, John Searle, Secret Societies, Sex Worship, Sitchin Studies, Smart Cards, Joseph Smith, Solar Power, Sovereignty, Space Travel, Spirituality, Stonehenge, Sumeria, Sun Myths, Symbolism, Tachyon Fields, Templars, Tesla, Theology, Time Travel, The Treasury, UFOs, Underground Bases, World Control, The World Grid, Zero Point Energy, and much more. Call **(800) 700-TREE** for our *FREE BOOK TREE CATALOG* or visit our website at www.thebooktree.com for more information.

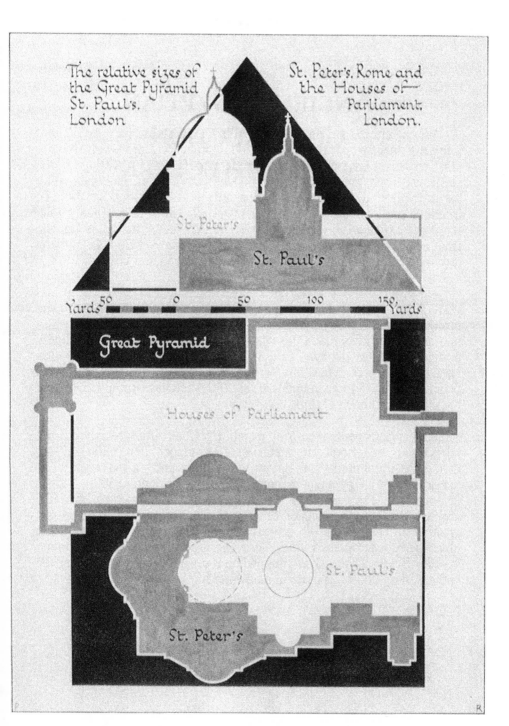

The relative sizes of the Great Pyramid St. Paul's, London

St. Peter's, Rome and the Houses of Parliament, London.

St. Peter's

St. Paul's

Yards 50   0   50   100   150 Yards

Great Pyramid

Houses of Parliament

St. Paul's

St. Peter's

FOREWARD

Out of all the books on Egypt, this one reveals much of the highly detailed methods the Egyptians used in building their stone constructions.

We often marvel at the incredible structures they built, not knowing exactly how it was done. The Great Pyramid's construction still remains a mystery to many people as far as the lifting of the stones to such great heights is concerned. To this day interesting theories abound, but no absolute proof, or a recreation of the true methods used, have ever been demonstrated. The authors openly state that their explanation of the Great Pyramid's construction is not the final answer and that many mysteries still remain, yet do a fascinating job in detailing how the huge stones were dressed and laid out.

On previous trips to Egypt I have been amazed at the deep and perfect drill holes found in some of the stones, not realizing that even though the ancient Egyptians did not have high speed drills, they still had and used drills that were adequate enough to do such absolutely perfect work—as depicted in a 5th Dynasty temple in Abusir (found later in the book). Many mysteries are cleared up in this incredibly detailed book, which includes over 250 sketches and pictures that back up the author's claims.

Any serious researcher into ancient Egyptian structures should not be without this book. It seems at first glance that many of the incredible monuments of Egypt were built and raised almost by magic. Touring Egypt is a magical experience for anyone with a sense of wonder. The structures are absolutely awe-inspiring, but the authors strip away much of the "magic" to reveal how hard work, creative ingenuity, and an advanced knowledge of mathematics and physics account for some of the amazing architectural feats performed in early Egypt. At the same time, when a mystery still remains for the authors, they do not hesitate in admitting it.

Paul Tice

PREFACE

THE purpose of this volume is to discuss some of the problems incident to the construction of a stone building in ancient Egypt. The material has been drawn partly from the architectural notes [1] made during the past thirty years by the late Mr. Somers Clarke, and partly from my own notes on the mechanical methods known to the Egyptians.

It might be imagined that the literature dealing with ancient Egyptian constructional methods would be very considerable, but this is far from the case. Though excellent works exist on architectural *style*, no work of any merit has yet appeared which discusses, in any detail, such prosaic subjects as the quarrying and dressing of a block, the function of mortar, the quality of the ancient foundations: in fact the successive steps taken by the old architects from the time the king ordered a temple until, with its surfaces dressed, sculptured, and painted, he dedicated it to the god. The more these various steps are studied, the more obvious it becomes that the Egyptian constructional methods differed radically from those of classical, medieval, or modern architects.

We do not pretend to have covered the subject exhaustively, and the following pages should rather be considered as merely a preliminary survey of a very wide field; indeed some chapters, such as those in which an inquiry is made into the methods of dressing the blocks, and into the explanation of the 'oblique joints', which are such a striking feature of the ancient masonry, are admittedly only tentative, and may well have to be modified in the light of future research. This applies equally to the problems of pyramid construction, where we are lamentably short of accurate data, owing to the fact that practically no pyramid prior to the Vth dynasty has been completely cleared of the debris surrounding it.

Sir Flinders Petrie, as far back as 1883, in a lecture given before the Anthropological Society,[2] pointed out, among other matters, many striking peculiarities in the ancient technique of stone dressing and cutting, and admitted that much was yet dark to him. It can almost be said that none of the features which puzzled him then has been properly explained since —at least in print. The reason is not far to seek; parties having concessions to excavate in Egypt have, in most cases, to take as a first consideration the

---

[1] These notes, contained in twenty-four architect's note-books, have been deposited in the Manuscript Department of the British Museum.   [2] See page 202.

necessity for acquiring objects of art for the societies which finance them, and in this matter pyramid and temple excavations are notoriously unproductive, and it is only within the last five years that the Egyptian Government has been able to provide funds for the excavation, under competent scholars, of the immensely important sites of Saqqâra and Dahshûr, which have already yielded a rich harvest.

The student with a knowledge of constructional work, engineering, or kindred subjects who desires to study the ancient building craft has hitherto been badly handicapped by lack of accurate data, not only as regards the details of the monuments, but on what is already definitely known concerning the methods and appliances used by the Egyptians, and on their knowledge of mathematics, astronomy, and other sciences. The result has been that many able architects and engineers have made statements concerning the manner in which blocks were transported and raised, and the monuments built, which they would never have written had the known evidence been available without a laborious delve into a host of works quite unknown to the general public. It is in this connexion that we hope the following pages may prove useful.

It is much to be regretted that, for some months before his death, my late friend and collaborator, owing to a stroke resulting in almost total deafness and blindness, was unable to assist in the final revision of his material and mine; indeed several chapters drawn almost entirely from my notes have had to go to press without his ever having been able to review their contents.[1]

We owe our warmest thanks to Mr. Cecil Firth and Mr. A. Lucas, of my Department, and to Dr. G. Reisner, Director of the Harvard Boston excavations, for valuable help and information, which is acknowledged as it is used; to Mr. J. Hewett, of the Survey of Egypt, for preparing many of our drawings for publication; to Prof. Sir Flinders Petrie for reading most of the volume when in manuscript and making many very valuable suggestions; and lastly to M. Pierre Lacau, the Director General, Antiquities Department, for permitting me to take and use many photographs of objects in the Cairo Museum and at Saqqâra.

R. E.

Cairo, 1929.

[1] Namely, Chapters II–V and VIII–X. Mr. Clarke was only able partially to revise with me Chapters II, III, IV, and VIII.

CONTENTS

LIST OF ILLUSTRATIONS

CHAPTER IV. TRANSPORT BARGES

CHAPTER V. PREPARATIONS BEFORE BUILDING

CHAPTER VI. FOUNDATIONS

CHAPTER VII. MORTAR

CHAPTER VIII. HANDLING THE BLOCKS

CHAPTER IX. DRESSING AND LAYING THE BLOCKS

CHAPTER X. PYRAMID CONSTRUCTION

CHAPTER XI.   PAVEMENTS AND COLUMN BASES

CHAPTER XII.   COLUMNS

CHAPTER XIII.   ARCHITRAVES, ROOFS, PROVISION AGAINST RAIN

CHAPTER XIV. DOORS AND DOORWAYS

CHAPTER XV. WINDOWS AND VENTILATION OPENINGS

CHAPTER XIX. BRICKWORK

APPENDIX I. EGYPTIAN TOOLS

APPENDIX II

INTRODUCTION

IN these days it has, happily, been realized that, to make a study of a country, whether ancient or modern, a great deal more must be done than describe its geography, catalogue its kings, discuss its internal politics, and relate how industriously it fell foul of its neighbours; we must not forget, for instance, the immense importance of being able to gauge the degree of civilization to which the inhabitants of the country may have risen.

The most satisfactory road leading to the knowledge of the civilization of a people is a study of its arts and crafts. The arts of Egypt, and many of its crafts, have been extensively studied by more or less competent scholars. The most striking of the latter, namely, the building craft, has, however, not received anything like the attention it deserves. This appears to be because very few persons possessing the requisite architectural and mechanical knowledge have sufficient leisure to devote to this study, since least of all branches of Egyptology can it be acquired otherwise than by personal examination of the many sites where the necessary information can be gathered—a scrap here and a scrap there.

Before we can completely account for the extreme conventionality—even monotony—of ancient Egyptian architecture, several factors have to be taken into consideration. For instance, it must be realized that the stone quarries were not open to the use of every one, at any rate until late times. The Egyptian world at large appears not to have been permitted to build with stone except in a very restricted manner. The quarrying and working-up of the material seem to have been in the hands of the state. It was natural, therefore, that when methods of work had once been established, the tendency to a hide-bound system, common to all bureaucracies, should develop itself and become crystallized—so thoroughly crystallized that we see, in Egypt, the same things done in the same way from the earliest dynasties down to the period of the Roman occupation, a matter of some 3,500 years.

There are other factors which must have contributed very much towards the stereotyping of methods, such as the geographical isolation of Egypt from other civilized nations who were keen builders, which deprived them of that healthy rivalry in architecture which had such a great effect on the western European nations. An even more important factor was the lack of variety in building material. The Egyptian had limestone in the north of his country and sandstone in the south, and both were worked and used by very much the same methods and required no radical difference of treatment. Granite and quartzite were also used, but sparingly, owing to the difficulty of quarrying and dressing. The only development of any moment in Egyptian architecture seems to have taken place when the sandstone quarries of Silsila were brought into general use, which enabled considerable spaces to be spanned and roofed with an easily worked stone.

In Egypt we observe comparatively few characteristics in architectural detail, or in the masonry, which can be described as local features, forced on the masons by the limitations of the available materials. How different is this from the infinite variety we find at home—a variety which has given birth to all manner of ingenious and beautiful methods of mastering the difficulties encountered. To take England as an example; in the eastern counties, stone is rare, but flint is abundant. The fertile invention of our forefathers quickly evolved from these materials that ingenious combination of flint-walling and stone dressings which gives such an individual character to the buildings in those parts of England. Again, in Somerset and the adjacent counties, where there is an abundance of excellent free-stone, we find a type of architectural detail and masonry perfectly suited to the material, but differing greatly from that of the eastern counties; it also differs from that evolved in Yorkshire, where the plentiful, but hard, stone made the working more laborious, but the result most noble and grave.

If we cease to compare district with district, but contrast the blossoming of medieval architecture in England and France, we can observe, amongst other differences, one which is most certainly due to the fact that it was easy in the Île de France to procure any quantity of excellent stone in blocks of considerable size. If we examine the masonry of most of the largest of the medieval churches in England, we are surprised to find how small are the blocks from which such vast structures are built. In many of the great French churches, on the other hand, the average size of a stone block is at least four times that of those used in the English churches. The Frenchman could, in consequence, safely undertake to build those structures of astounding hardihood in conception and execution which, with us, were not even contemplated.

The last factor which plays a large part in the development of architecture in most countries is the presence and condition of roads. For instance, in England, during the blossoming of church building which began soon after the Conquest, we find a people ambitious to put up great buildings, but without any roads on which they could carry their materials, with the result that only blocks of insignificant size could be used. As time went on, waterways were improved and roads—of a sort—developed. We then find that the masonry improves, and that stone is brought from greater distances. Consequently, having a larger command of materials, the masons could venture on and master difficulties which at first had been quite beyond their powers. In Egypt there was no such progress; the Nile was a good road before the First Dynasty and is still so to-day.

The more we know of the building craft of the ancient Egyptians, the quarrying, the stone-dressing, the masonry of walls, pylons, pyramids, and columns, the more are we amazed at the strange contradictions in their character which are revealed. We cannot help admitting that they were perhaps the best organizers of human labour the world has ever seen, and their method of carrying out a task always appears to be the most efficient and economical, in principle at any rate, when we take into account the appliances which they knew and the methods of transport at their disposal. Their powers of transport by water were astonishing; whether thousands of blocks were required for a temple or whether a single block weighing 1,000 tons had to be brought, their boat-building powers were fully equal to the demand made upon them. Some of their masonry has never been excelled for fineness of jointing, although the blocks may weigh up to 15 tons apiece, and we look with even greater respect on the giant structures they erected when we discover that the only mechanical appliances they knew were the lever, the roller, and the use of vast embankments. On the other hand, we are often equally astonished at their short-sightedness in matters which we now consider of primary importance, such as the necessity for 'breaking joint' and the value of good foundations. Their neglect in the matter of foundations is strikingly exemplified in the series of mortuary temples of the XVIIth and XIXth dynasties at Thebes. As soon as one had fallen into disrepair owing to bad masons' work, neglect in foundations, or overlooking the gradual rise of the Nile bed, the reigning king took the blocks of the older temple (unless he had a special regard for the king commemorated therein) and his architects committed exactly the same faults in masonry, foundations, and choice of site as their predecessors, ignoring completely the examples before them of what would in a few years be the inevitable fate of the new building.[1] The Egyptians' pro-

---

[1] It has been suggested that the kings may not have cared overmuch how long the buildings lasted, once the meritorious act of building and dedicating them had been performed. This can, however, hardly be reconciled with the fact that the Theban mortuary temples were apparently constructed for the express purpose of maintaining the worship of the dead king, who ranked as a god.

pensity to 'spoil the ship for a ha'p'orth of tar' can be noticed at almost
every stage of their work.[1] Although their walls usually consisted of but
two skins of masonry, with the space between them filled in with rough
blocks or even rubble, they frequently—at any rate in the New Kingdom
—used superfluous half-drums of columns in the facing, adding considerably to the weakness of a form of wall which, at best, was none too rigid.[2]
The collapse of the Ramesseum, where half-drums were frequently used,
was brought about almost entirely by faulty work of this kind. We can see
this false economy in the very last stages of the building, namely, in the
placing of the architraves and the roof-slabs, where any weakness would
be expected to be specially guarded against.

Egyptian masonry rose to a peak of excellence during the reign of King
Khufu, after which no advance in methods of construction was made,
though new forms of architecture developed. The quality of the masonry,
broadly speaking, steadily deteriorated.

[1] See Figs. 127, 168 and 169, and 174.   [2] This fault can also be seen in the Colosseum at Rome.

I

THE EARLIEST EGYPTIAN MASONRY

IN a country of so uniform a character as Egypt, it might well be expected that it would be a simple matter to discover the prototypes of the architectural forms met with in stone buildings. As a matter of fact, however, one has to be very cautious in committing oneself too definitely by insisting that any single form of primitive reed-and-mud construction of which the people in earliest times might be expected to have made use contains the germ of all the architectural forms found, for example, in a pylon. Several authors have brought forward more or less ingenious suggestions to explain Egyptian architectural forms,[1] but none of them can be said to be completely satisfactory.

It seems likely, indeed almost certain, that both brickwork and reed-and-mud constructions played their part in the evolution of Egyptian forms; it has even been suggested that the panelling seen in the mastabas had its origin in portable wooden huts (p. 214), though this extremely ingenious theory is not universally accepted.

It is not without interest to consider what the most primitive form of house may have been like. Nowadays, though in his village the Egyptian peasant lives in a brick hut, when he has to spend any considerable time in the fields he makes a shanty of maize-stalks (Arab. *bûṣ*). His method of constructing such a shanty is to lash the reeds, on the ground, into, as it were, a large mat, by means of palm-ropes. These mats, when set upright, form the walls of the hut. Sometimes the bottoms of the reeds are buried a few inches deep; at other times they merely stand on the surface of the ground. To give the structure more rigidity, bundles of maize stalks about three inches thick are often lashed horizontally near the top of the wall and vertically at the corners. It has been held that this is the prototype of the torus mouldings and rolls which, in Egyptian buildings, are usually represented with lashings round them.

In ancient Egypt, maize-stalks were not known, and one of the commonest reeds seems to have been the papyrus. This plant has a peculiar head, which was freely imitated in stone columns and depicted in tomb-scenes. If we assume that the papyrus plants were used with the heads left

[1] PETRIE, *Arts and Crafts*, p. 63.

on, the effect of a line of them forming the wall of a hut or enclosure would resemble in no small degree a cornice. In the curve of many Egyptian stone cornices we have, however, a form of decoration which cannot, apparently, be referred to a reed construction; often it strongly suggests the palm-frond.

It has therefore been suggested[1] that the Egyptian cornice and torus moulding took their rise from a primitive form of hut constructed of palm-fronds interlaced in a framework of poles. Some confirmation is found for this suggestion in the representations of certain shrines on the walls of tombs and temples. In a scene of the shrine of a lion-god on a block found at Memphis,[2] the face of the shrine is covered with a criss-cross pattern, and similar examples are known elsewhere. It must be admitted, however, that, in modern times, the peasant makes very little use of the palm-frond for the walls of his shanty, though such palm-huts are not unknown. Other suggestions have been brought forward to account for the origin of the cornice and torus roll, but a discussion of them is somewhat outside the scope of this volume.

It might well be imagined that the pyramid could justly be considered to be the direct descendant of a conical pile of stones placed over the grave of a primitive chief, but the evidence is strongly against this being the case. The earliest superstructures known[3] are not conical at all, but rounded, and it is clear that the pyramid grew, by distinct stages, out of the low platform, square in plan and with a batter on its sides, which formed a superstructure for the early Old Kingdom tombs. A pyramid, in fact, is the development of a compound 'mastaba'.[4]

Columns can be referred back in many cases to extremely primitive forms. A bundle of reeds, when suitably worked up with mud, can be made to support such heavy weights as the counterpoise of the water-raising appliance now known as the *shadûf*. Another and stronger supporting medium is the palm-trunk, which is of no use for planks. It requires no great stretch of the imagination to furnish a column with a head suggestive of the material from which it has been derived; thus, at Abusîr, columns of the Vth dynasty are found with palm-frond capitals, while the papyrus column appears to be of even earlier date, since the capital of a pilaster of this form is found at Saqqâra dated to the IIIrd dynasty (Fig. 7).

[1] PETRIE, *Arts and Crafts*, p. 63.
[2] PETRIE, *The Palace of Apries (Memphis II)*, Pl. XVIII.
[3] PETRIE, *Gizeh and Rifeh*, Pl. V E.
[4] The superstructures of ancient tombs, rectangular in plan and with a batter on each face, have long been known among archaeologists under this name. It is the Arabic word for the mud platform used as a seat outside the doors of the houses.

The next step in the decoration of columns seems to have been to furnish them with floral capitals suggestive of the lotus or lily, which it was the custom to tie round the tops of the posts supporting the roofs of houses. The Lotus Column comes into use as early as the Vth dynasty (Fig. 159).

A hut, in Egypt, is roofed with reeds or palm-fronds, which keep out the heat admirably. There is no evidence to show that planks were ever used for this purpose. It would not be justifiable, therefore, to suppose that the flat roof-blocks of the temples had their origin in planks; it is more probable that they developed naturally from the use of good stone.

The earliest stone roofs known are of blocks laid on edge, the depth being at least twice the breadth. The under sides of these blocks are cut into a semicircle, often painted to represent logs or palm-trunks laid side by side. This form of roofing is seen in some of the IIIrd and IVth dynasty mastabas at Saqqâra and Gîza, and it also appears in the newly discovered chapels of the time of King Zoser. Such a series of blocks has the disadvantage of offering many opportunities for rain to enter, though it must be admitted that, in the Zoser masonry, the good condition of the paint on the under side of these stone 'logs' shows that very little rain actually did come through. When the increasing use of sandstone made it possible to roof comparatively large spaces with slabs rather than with blocks of the form just described, it seems that the tradition of the log-roof almost fell into oblivion.

Having very briefly surveyed the manner in which some of the more important architectural forms may have taken their rise, we have next to inquire into the actual birth of masonry in Egypt. Until recently this seemed fairly clear. At Abydos, in the tomb of King Kha'sekhemui of the IInd dynasty,[1] the tomb-chamber was lined with stone instead of having a lining of wood like those of his predecessors. The tomb-chamber measured 17 feet by 10 feet and was nearly 6 feet in depth. Besides this building, a granite jamb for the temple gateway was found at Hieraconpolis.[2] Sa-nakht, the first king of the IIIrd dynasty, had a mastaba at Beit Khallâf, 200 feet long by 80 feet wide, which contained three small stone chambers, and at Beit Khallâf there is also a large mastaba, 300 feet long and 150 feet wide and over 30 feet high, which some consider to have been one of the tombs of King Zoser. In it there is a long descending passage, barred by five great portcullis blocks moving in masonry, leading to a wide horizontal passage. At the bottom there are a dozen chambers over 50 feet underground. From clay sealings, there is no doubt that it must be dated

---

[1] PETRIE, *History of Egypt* (1923), i, p. 37, and *Royal Tombs*, ii, pp. 12–14.

[2] QUIBELL, *Hieraconpolis*, i, Pl. II. Now in the Cairo Museum.

at any rate to Zoser's reign.[1] The next example of masonry known was the Step Pyramid of Zoser; this, for size, far exceeded anything which went before it, though all that could be seen of it until recently was the rather coarse core. It contained peculiar decoration, such as a chamber beautifully lined with glazed tiles, on which account some scholars considered that much of it had been reconstructed in late times. The IIIrd dynasty was followed by a period, believed to be of about 100 years, during which five kings reigned, of whom we know nothing save the names.[2] The so-called False Pyramid of Sneferu, the last king of the IIIrd dynasty or the first of the IVth dynasty, was the next piece of masonry known, and its fine casing and well-built pyramid-temple are nearly on a level in quality with the pyramids at Gîza which followed it.

This, then, was the supposed sequence leading up to the gigantic constructions undertaken in the IVth dynasty. The recent researches carried out by the Antiquities Department at Saqqâra have, however, compelled scholars to revise their views very considerably on the birth of masonry in Egypt, for the buildings around the Step Pyramid, also certainly of the IIIrd dynasty, are of a type of masonry never hitherto encountered, and show architectural features which are quite new to archaeologists.

A superficial examination of the masonry of King Zoser's reign might well give the impression that, apart from its delicacy and its pleasing architectural forms, it is of superior quality to that of the pyramids and temples that followed it, and the idea seems to be gaining ground that this form of masonry became, for some mysterious reason, a lost art. This is entirely erroneous. The Zoser masonry is, generally speaking, of much poorer quality than that of good mastaba and pyramid masonry of the IVth and Vth dynasties, and the structures, owing to the smallness of the blocks used, were not calculated to last any great time. Either during the end of the reign of Zoser or in the little-known period which followed it, greater strength was sought, especially in the royal funerary buildings, and the size of the building-blocks was greatly increased. As soon as the weight of the blocks became such that they could not be *lifted* by a party of men, considerable changes in the dressing and laying technique had to be evolved, since the Egyptian appears never to have invented the pulley and lifting-tackle (p. 85). The more the IIIrd dynasty small-block masonry is studied, the more clear it becomes that the megalithic masonry which followed is merely a development from it.

---

[1] Garstang, *Mahâsna and Bêt Khallâf*, Pls. VI & VII.

[2] The Abydos list only gives three kings in this period. Petrie, *History of Egypt* (1923), i, p. 39.

Fig. 1. Part of the façade of the chapel of a princess of the IIIrd dynasty. Saqqâra. (Photograph by the Antiquities Department, Egyptian Government)

Fig. 2. Fluted columnoids in the small temple behind the *sed*-festival temple at Saqqâra. IIIrd dynasty. (Photograph by the Antiquities Department, Egyptian Government)

Fig. 3. Colonnade of King Zoser at the south-east corner of the temenos wall round the Step Pyramid. IIIrd dynasty; Saqqâra. (Photograph by the Antiquities Department, Egyptian Government)

Fig. 4. Cross hall at west end of Zoser's colonnade. IIIrd dynasty; Saqqâra. (Photograph by the Antiquities Department, Egyptian Government)

The great surprise which Zoser has provided for the student has been the use of pilasters, though the presence of flutings on some of them created more interest among the general public. Flutings occur sporadically almost throughout the dynasties (p. 139), though pilasters are hardly known even at such late times as those of the Ptolemaic and Roman temples of Edfu, Qalabsha, and Dendera. A pilaster may be described as a 'ghost of a column', and has little, if any, constructive value; it belongs to the masonry of a wall, but makes it no stronger. In the masonry of Zoser, the strange anomaly is found of the presence of the pilaster but not of the free-standing column. Apart from their decoration, they are of two types, the true pilaster (Fig. 1) and what may architecturally be described as a pilaster, but must really be considered as the incorporation of a wall and a column (Figs. 2–4). Mr. Cecil Firth, describing these in the *Annales du Service*, vol. xxv, p. 158, remarks: 'It is quite clear that the builder was perfectly acquainted with the free-standing column, but that in this case he preferred to carve it at the end of a short wall in order to support the heavy roof of limestone beams painted red to represent logs of wood . . .'

Dr. G. Reisner, referring to the Gîza mastabas of the IVth dynasty, has kindly informed one of the writers that:

'slabs of limestone, of an extreme length of 250 to 285 cm., formed a practical roofing material; the longer slabs are from 50 to 80 cm. thick. The limestone in the walls lies in its natural position with the strata horizontal, as also do the roofing-slabs and the architraves. The square pillars used in the fourth dynasty have the natural strata running vertically. The stone is so well selected and the architrave so carefully set that the splintering of a column is very rare indeed. The strain on the middle part of the architrave and the roofing-slab was not more than the stone could bear. The span over which the weight was borne was usually between 120 cm. and 150 cm. and over these roofs there was usually only a layer of filling 20 to 100 cm. thick.'

It seems that Zoser's architects could trust the stone to span a gap of some nine feet, but did not dare to put roofing-slabs on an architrave spanning this distance; on the other hand, it may have been that it was the supporting power of free-standing columns constructed of small blocks that they doubted. It must not be imagined that at this period the actual use of logs for roofing was merely a tradition from the dim past. Dr. Reisner has informed the writers that the northern corridor and the magazines of the Mycerinos temple at Gîza were roofed with wooden logs.[1]

[1] In the great IIIrd dynasty mastaba on the north of the Step Pyramid, baulks of wood were placed along the top of the masonry walls lining the corri- dors, and on these the roofing beams were laid. In this building the under sides of the roof-blocks were rounded to represent logs. The purpose of the

Among the more interesting architectural forms found in the Zoser masonry may be cited the ribbed columns which formed the colonnade on the south-east corner of the girdle-wall of the Step Pyramid (Figs. 3 & 4). This form of column is never subsequently encountered. Its capital was a simple one, surmounted by a wide abacus.[1] Another strange form is the papyrus-stalk pilaster, with the triangular section which is characteristic of the reed (Fig. 3 & Fig. 153). The capital represents the spreading papyrus flower, and this form continued to be used, on columns, until the latest dynastic times. The façade of the upper part of one of the buildings was originally ornamented with a series of slender fluted pilasters —three to each chapel—supporting a curved or arched cornice. These pilasters, so far as is yet known, terminated above in a form resembling two pendant leaves, one on each side (Figs. 5 & 6), which were pierced from front to back with a roughly cylindrical hole, perhaps to assist in the suspension of an awning for sheltering the unroofed court in front. They may, on the other hand, have received copper spouts for draining the roof.

An interesting problem arises in connexion with the Zoser masonry, whether the architectural forms and technique developed during the reign of Zoser or whether they had a considerable history behind them. At first it seems incredible that they could have been evolved so quickly, and the presence of the pilaster has been held to be proof absolute that free-standing columns must have existed in earlier stone buildings. The writers are inclined to believe that the art of laying finely dressed blocks may well have developed during Zoser's reign, the forms being translated from brick and from vegetable growths. Free-standing columns must, indeed, have been known before Zoser's time, but perhaps only in the primitive form of palm- and tree-trunks and reeds stiffened with mud (p. 6). Such columns, when translated into stone, had, it seems, to be strengthened by combining them with the walls, and from this the pilasters in the chapels of the IIIrd dynasty princesses may have at once developed. In other countries, where free-standing columns seem to have been constructed from the outset, the pilaster did not develop for a long time afterwards. It is possible that free-standing stone columns may have been tried by Zoser's architects, but that they were not found to be sufficiently strong to bear architraves and roof-blocks. It is too early to be certain on these points, and future excavations at Saqqâra may provide evidences of free-standing columns. Even in the chapels and colonnades already cleared, traces of progress can be observed;

baulks was to neutralize any unevenness of pressure which the roof blocks might exert on the wall.   [1] See LAUER, *Annales du Service*, vol. xxvii, pp. 112-33, where a restoration of the capital is shown.

Fig. 5. Head of a pilaster. Masonry of King Zoser; IIIrd dynasty; Saqqâra. (Photograph by Antiquities Department, Egyptian Government)

Fig. 6. Diagonal elevation of the head of the pilaster shown in Fig. 5

Fig. 7. Pilaster imitating the head and stalk of a papyrus plant, from the chapel of a IIIrd dynasty princess of Saqqâra. (Photograph by the

the courses on the masonry become more and more straight, and other tricks of the craft enabling closer joints to be constructed can be seen in their growth. Building with more or less rough stone blocks had been known in Egypt for some centuries before Zoser, and the knowledge of cutting and dressing the hardest rocks had been well known from early predynastic times, so it is less surprising that, when once builders had conceived the idea of copying the attractive architectural forms seen in the houses in such a fine, easily worked stone as was found in the Tura-Ma'sara quarries, the art should have developed with great rapidity. When gigantic structures built of great blocks were constructed, intended to last for all time, the delicate decoration of Zoser was no longer used, since it would have been out of place in a building where grandeur and extreme accuracy of work were the objects of the architects.

II

QUARRYING: SOFT ROCKS

ROCKS may, for convenience in studying the ancient methods of quarrying, be divided into 'soft' and 'hard'. Entirely different methods of 'getting' the stone were used for the one and for the other. The soft rocks which were used for building are the limestones and the sandstones and, to a much smaller extent, the calcite, or Egyptian alabaster. The hard rocks are the granites, the basalts, the diorites, and the quartzites.

Limestone extends from Cairo up the river as far as Esna, where it gives way to sandstone, which, with occasional outcrops of granite and diorite, extends throughout Nubia.

From the earliest times, the principal limestone quarries were those of Tura and Ma'sara, which lie about four miles south of Cairo. Another interesting series extends from Beni Hasan to Sheykh 'Abâda; in fact, wherever a stratum of good building-stone occurs anywhere near the Nile, there may be seen the ancient workings (Fig. 8). Since stone is required to-day for the construction of embankments to check the encroachments of the Nile, a certain number of ancient quarry-faces have had to be sacrificed, though the Antiquities Department endeavours to respect as many of the ancient workings as possible. This sometimes involves great expense, as, for example, in the case of the Beni Hasan quarries, where stone has to be brought from a considerable distance up river in order to keep the ancient Egyptian quarries intact. It is a great pity that no 'learned society' sends out a qualified person to make a complete study of the ancient quarries. The reason for this neglect appears to be that such a study would not be likely to furnish objects of interest to museums, to the securing of which new information is too often a secondary consideration.

During the period in which limestone was the principal building-stone —the Old and apparently the Middle Kingdoms—buildings are conspicuous by the relative smallness of their apartments. Limestone is not the medium for architraves; the most that can be spanned, for instance, by Tura or Ma'sara limestone is about nine feet. Even when such a space is spanned by an architrave, it will not bear roof-blocks with any likelihood of lasting. It was necessary, in the early masonry at Gîza, to obtain granite for the purposes of roofing if any apartment of considerable size was required, or else to construct a corbelled or a pent roof (Figs. 218 & 219).

Fig. 8. Ancient limestone quarry. Beni Hasan

Fig. 9. End of a gallery: Ma'sara quarries

Fig. 10. Top of a quarry-face in a gallery at Ma'sara, showing where the stone has been cut to waste at the roof to 'get' a new series of blocks. The slots on the right were to enable the quarrymen to climb up to their work

Fig. 11. Entrances to ancient galleries in the limestone quarries at Ma'sara. Note the figure standing before the central opening

It was due to the exploitation of the fine sandstone quarries at Silsila that the New Kingdom architects were able to build on such a magnificent scale. We are very short of information on Old Kingdom temples in Upper Egypt, but we may safely assume that, had the value of sandstone for roofing been known, it would surely have been brought to Gîza, and that the architecture would have been very different from what it actually was. The same remarks apply to the Middle Kingdom, though our knowledge of this period is practically confined to the XIth dynasty temple at El-Deir el-Bahari and to the Labyrinth at Hawâra, neither of which shows apartments of any great size.

All the soft stone blocks were quarried on the same general principles; the vertical faces of the blocks were cut out by means of metal tools and the blocks were finally detached from below by the action of wedges. The most striking point to be observed in ancient Egyptian quarries is the orderliness with which the stone was 'got'. Instead of being wrenched from the hillside, as is so often done to-day, in ancient times it was removed in roughly rectangular blocks.

We may, for convenience, divide the soft stone quarries into 'open' and 'covered'. The open type is found where the good stone starts from the surface, as at Beni Hasan, Silsila, and Qertassi, and the covered type where it only lies in a stratum at a considerable depth below the surface, as in many of the Tura and Ma'sara quarries (Fig. 9). Though the principle used in detaching a block was the same for both, the order in which the blocks were got was sometimes considerably different.

In a covered quarry, the faces were nearly always maintained vertical, and worked as far down as possible before a new set of blocks was extracted from the face behind. One vertical face having been finished, the rock was cut to waste at the roof of the gallery as far in as was required to get behind what was to be a new series of blocks, a kind of shelf being made, of a sufficient height—usually a little over a yard—for a man to kneel or squat in it and make a vertical cut at the back of the line of blocks, and other cuts to separate one block from its neighbours. This line was removed by driving small wedges in from the front, thus lifting the block from its base. Slots up the face of the quarry are often seen, by which the ancient quarrymen climbed up to their work (Fig. 10). It will be noticed, on the quarry faces shown, that no traces of wedges can be seen. It is only when a quarry is excavated until the lowest workings of any particular face are reached, that it is possible to observe the wedge-marks on the rock from above which a block has been lifted. In the case of the Tura limestone the rock parts with great ease; in modern quarrying, two very small slots are cut at the base of the block, and a few taps with a sledge-hammer on chisels inserted into these easily liberate it.

In the Tura and Ma'sara quarries, laminations are not very marked in the best strata, but in the quarries in which they occur, such as those at Silsila, they are freely taken advantage of.

Many of the quarries had sloping ways leading down from them, some-times very steep, possibly so that the blocks could be rolled down to the sleds on which they were transported (p. 88). Since the blocks were not fine-dressed at the quarry, such rough treatment would not cause any serious damage.

Few things are more impressive than the large covered quarries. At Tura and Ma'sara, for example, they appear from the river as almost rectangular openings—some like great doorways, others wider than they are high (Fig. 11). The entrances appear as dense black spots against the intense bright-ness of the sunlit cliffs. From the floor-levels of the openings, long shoots of rubbish may be traced descending into the plain below. On approaching one of the openings, one begins to see that within there are massive pillars, more or less square in plan, but very irregularly placed. They support the overlying strata, in which the rock is not of sufficiently good quality for building. The opening, which appears of insignificant size from a distance, is often in reality more than twenty feet high, and the gallery sometimes goes hundreds of yards into the mountain, pillar following pillar until they are lost in pitch blackness. Herodotus states (Book II, sect. 24) that the quarries which supplied stone for the Gîza pyramids were on the east bank of the Nile, but such a statement should be accepted with some caution, since, like so many ancient authors, he did not sift his evidence with over-much care. It is almost certain, however, that the stone for the *casing* of the pyramids and the fine lining of the countless mastabas which extend from Abu Rawâsh to Saqqâra came from the Muqattam-Tura-Ma'sara area. The limestone from this region was not only sent freely to the great Delta cities, such as Sais, Tanis, Buto, Bubastis, and Mendes, but it was even shipped up-river for door-jambs and lintels as far south as Aswân. The drain on the quarries during the three thousand odd years they were in use is almost past imagination, and the appearance of the cliffs in their original state can hardly be gauged.

The open quarries, of which Silsila and Beni Hasan are the finest examples, are hardly less striking than the closed ones. At Silsila, the Nile flows through a narrow gorge without any cultivation on either side of it. For half a mile along each bank, huge bays—some almost like courtyards—can be seen one after another on both banks of the river, the quarry-faces being almost vertical, and showing the same orderliness of work which seems always to have characterized ancient Egyptian methods. The quarries are open to the sky, since the good stone extends downwards almost from the surface. The cleavage of the stone is horizontal, and at intervals there are vertical faults extending far down into the ground, and quite unstained. Some of the faces are as much as forty feet high. Unlike the Tura and Ma'sara quarries, those at Silsila show few evidences of tunnelling.

The Beni Hasan quarries are not worked in the form of closed bays, but otherwise the stone seems to have been extracted by methods exactly similar to those used at Silsila. They extend for at least three miles along the cliffs in two very definite strata, away from which the stone is powdery and poor. The best stone is an extremely hard limestone full of little fossils.

South of Silsila there are many sandstone quarries, but none of any magnitude until the First Cataract has been passed and Qertassi is reached. From here came the stone used for Philae, with its temples, quays, and colonnades, and for the temple of Qertassi itself. It is obvious that a very great amount of stone has been taken from this place—far more than would be required for Philae alone. It must, therefore, have been exported for other temples, but it is difficult to determine for which, since the Nubian temples all had their quarries close to them, and it can hardly be assumed that the stone was transported across the First Cataract.

Proceeding south from Qertassi, the quality of the sandstone becomes, generally speaking, worse and worse, and the temples constructed from the local quarries have suffered in consequence. At Soleb, for instance, between the Second and Third Cataract, there is the ruin of a stately and ambitious temple of Amenophis III, closely resembling in its architectural treatment the temple of that king at Luxor. Constructed as it is of the miserable Nubian sandstone, the architect must have met with the greatest difficulties; the stone was quite unsuited to the conventional trabeated design on a large scale, and it must have given way very early in the history of the structure. At Gebel Barkal, the material was, if possible, even worse, and the temples are consequently in a deplorable state of ruin.

The chief difference between 'getting' the stone in a covered and in an open quarry is that, in the latter, a larger number of blocks at one level can be worked on simultaneously, whereas in the former the part cut to waste below the roof rarely extends inwards for more than three blocks' breadth. When work in an open quarry has been stopped, the floor presents the appearance of steps (Fig. 12). In such quarries, the tops of the blocks to be 'got' were often marked out by a succession of indentations with a chisel or with red ochre, the double line so indicated—that is, the top of the next separating trench—being some 4½ inches wide. In the Silsila quarries, there is a block that has not yet been detached from its bed, but is in all other respects ready for removal. It measures 20 feet by 2 feet 7 inches by 5 feet high and was intended, no doubt, for one of the massive architraves

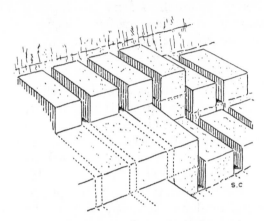

Fig. 12. Diagram showing method of extraction
of blocks in a quarry.

which extend from column to column to carry the roof-slabs. Only two ends of the block had to be cut, since advantage was taken of the vertical faults which are a feature of these quarries.

At Silsila, the blocks were removed in the usual manner by the action of wedges driven in horizontally at the base, the distance between successive wedge-slots being some 4½ inches. Occasionally, traces of wedges which have been used vertically may be seen, where a block has been torn from the quarry-face instead of from its bed. Wedges acting vertically seem never to have been used in the Ma'sara quarries.

Another interesting example of ancient quarrying is seen in the area north of the Second Pyramid, whence large blocks have been removed to leave this side of the pyramid level and to fill up the low parts on the southern side. Even in this 'cutting and filling' work, the same economy in extraction can be observed (Fig. 13). The bottoms of the separating trenches, which measure some two feet in width, can still be seen over a large area, the blocks which were removed having been, on the average, nine

Fig. 13. Levelled area on the north side of the Second Pyramid at Gîza. The blocks removed from here, measuring nine feet square, were used to fill up the low ground on the southern side

Fig. 14. Tool-marks on a quarry-face near Beni Hasan

Fig. 16. Closer view of the tool-marks in the separating-trench in the Qâu quarry (*Fig.* 15)

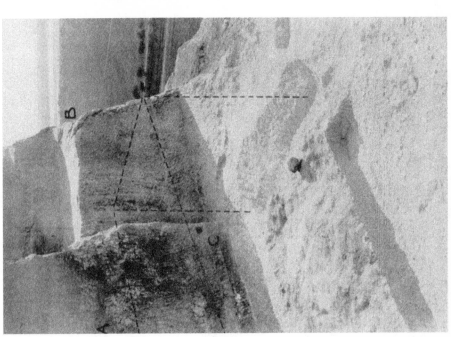

Fig. 15. Quarry at Qâu, from which a large limestone block has been removed. (Photograph by Guy Brunton, Esq.)

nine feet square. Whether, at any period during the working of this quarry, it ever had the appearance of steps (Fig. 12) is uncertain; it seems likely, however, that the blocks were removed layer by layer, since it was required to leave this area level, and such would be the most practical method.

In ancient Egyptian quarrying, it is remarkable how narrow are the cuttings used for separating a block from the parent rock or from its neighbour. Except for the largest blocks, the cuts rarely exceed 4½ inches in width. The difficulty of making a cut with a chisel of any depth is by no means trifling. A plain copper chisel (Fig. 263) struck with a wooden mallet, seems to have been the normal method of cutting the soft rocks, and if this was the method used at the bottoms of some of the separating trenches between successive blocks, the chisels would have had to be of considerable length. A very fine copper chisel, measuring 19 inches in length (Fig. 263) is known from Gebelein, but unfortunately its date is uncertain, though it is undoubtedly pre-Roman. In spite of the existence of such long chisels, it has, however, seriously to be considered whether some form of mason's pick can have been known to the Egyptians. To-day, in the Tura quarries, stone is cut with a pick, which is a hammer with a steel head of some five pounds weight pointed at both ends, the haft varying from one foot to two feet in length. This is held in both hands and the stone is struck with fairly soft blows. The marks left on the quarry-face after extracting blocks with this tool very strongly resemble those left by the ancient quarrymen. The work with a mason's pick leaves cuts lying more or less on a curve, whereas those made by a chisel and mallet are generally either straight or quite irregular. Many of the Egyptian quarry-faces show the cuts distinctly lying on the arc of a circle (Fig. 14), but this is not sufficient proof that the Egyptians knew the mason's pick of the type used at the present time; certainly no example has been found, nor is it depicted in the sculptures. It is quite within the bounds of possibility, however, that some form of pointed tool was attached adzewise to a haft for use in quarrying. At any rate, the adze is amply vouched for in ancient times, and is known to have been used, in the late dynasties at least, for stone dressing. If such a tool were used by the Egyptians, it would have had to be fairly heavy, or else weighted in some manner, since the value of the mason's pick for cutting stone depends on its mass rather than on the force with which it is brought down. The problem of the forms of quarrying tools used for soft rocks has to be left in this rather unsatisfactory state until further evidence is forthcoming.

The nature of the metal of the stone-cutting tools also presents problems which have not yet been satisfactorily solved. This, however, applies to the attack on the hard rocks rather than the soft, and is discussed in Chapter III. There is little reason to believe that the tools were of other metal than copper or bronze, which will cut the softer rocks with comparative ease if tempered by hammering and heating, though admittedly at the expense of constant re-sharpening.

In quarrying a block of any great size, as, for example, one for a large sarcophagus, the separating trenches would have, of necessity, to be at least two feet wide to enable a man to get into them and work, and to take the points of large levers by means of which the block could be handled (p. 88). The British School of Archaeology in Egypt, during their season at Qâu, kindly cleared, for one of the writers, a small quarry from which a block of such a size had obviously been extracted (Fig. 15). The rock in this quarry is a very hard limestone. The quarrying of the higher level (above AB) has been done by pounding it with balls of hard stone, probably dolerite, the normal method for the hard rocks (p. 26), while a large block, which measured about 9 feet by 5 feet by 5 feet deep has been cut out by means of metal tools. In the illustration, the probable dimensions of the block extracted are indicated by dotted lines. The tool-marks on the sides of the separating trenches are almost parallel, and rarely pass into one another (Fig. 16); this suggests that the tool was driven into the stone almost vertically, that small trenches were cut across the line of the separating trench at intervals of from one to two inches, and that the stone between these trenches was chipped or hammered away. It must be admitted that to do this would require a tool of great hardness.

It has been remarked that the blocks were detached from their beds by means of wedges, and it is of interest to determine whether they were small pieces of metal hammered into the slots cut for them, either with or without the two metal strips on either side of the wedge ('plug') now known as 'feathers', or whether the wedge was made of wood and caused to expand by being wetted with water. For the smaller limestone blocks it is certain that hammered wedges were employed, quite possibly, in some cases, in conjunction with 'feathers'. Wedges of iron are actually known from excavations in the Ramesseum, but they are of late dynastic date.[1] 'Feathers' are also known in late Egyptian times, though no examples of early date have been preserved. In the quarries of the Wady Hammamât there are frequent references to workers of iron, who may well have been those who

[1] For further notes on wedges, see p. 23.

forged the wedges. In the Qâu quarry already described, it will be seen (Fig. 15) that below, where the block has been removed, there are trenches which had been cut partly beneath it. On the surface from above which the block was torn there are no traces of wedge-slots, from which it may be concluded that the trenches played some part in the extraction. By no possibility could any form of hammered wedge have been used in this case, though wood packed into the trench and made to expand by wetting might well have been the medium employed to tear the block from its base. In the granite quarries of Aswân, wedge-marks are sometimes seen

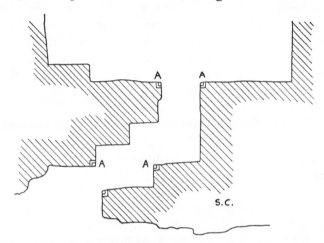

Fig. 17. Sketch-plan of the entrance passage to one of the enclosed bays in the ancient sandstone quarries of Gebel Silsila.

of so great a size that, without the Qâu example, it might be deduced that expanding wooden wedges had been employed.

Apart from the problem of the methods employed in extracting the blocks, a study of the quarries reveals many features difficult of explanation, especially when the floor of the quarry is not clear of debris, which is the case in the vast majority of them. In the Silsila quarries, for instance, no satisfactory explanation is forthcoming of the reason why they were left in the form of enclosed bays, the more so since the passage between the bay and the river is rarely straight, sometimes, on the contrary, having quite a tortuous course. Since the quarries are quite close to the river, it is a possibility that the descent was found to be too steep, and that the quarry-men feared that a block would be difficult to control down a straight path. The floors of such of these passages as have been examined are quite rough, and there are no indications that they had been prepared for sleds (p. 89). At certain salient points in these zigzag entrances to the quarries (Fig. 17, *A*), holes have been cut in the angles of the rock. These holes are more or less square in section and measure about 12 inches in height and some 6 inches in width. A rope passed through them might conceivably have assisted in handling the blocks, but it must be admitted that the holes show no signs of the friction which a rope so used would be expected to produce. Similar piercings can be seen at the water's edge. These holes are not only known in the Silsila quarries, but also occur here and there at Tura and Ma'sara. In the latter quarries a pair of holes of the same dimensions as those cited can be seen about 15 feet above the present ground level at the entrance to one of the galleries (Fig. 18). No explanation of their use is at present forthcoming. In the Qâu quarry, on the other hand, the hole at *A* (Fig. 15) can be explained; its undoubted function was to serve for the attachment of a rope to enable the quarrymen to clamber up on to the platform from the slope leading to the cultivation, in order to save a detour of some twenty yards. It is more prudent to leave the explanation of these holes until the floor of the quarry near them is cleared, when some such connexion as that observed in the Qâu quarry may well be forthcoming.

Alabaster (calcite), being a soft rock, was quarried in much the same manner as the limestones and sandstones. The best known of these quarries, the *Het-Nub* of the ancient inscriptions, lies out in the desert about 15 miles south-east of El-'Amârna. In form it is a gigantic pit cut down into the rock to a depth of more than 60 feet and approached by a narrow sloping way (Fig. 19). The whole excavation measures about 200 yards across. On the sides of the entrance passage are the names of some of the overseers who were responsible for getting the stone and those of the kings who sent them, some of which date back to the IVth dynasty. The quarry-faces are constantly intersected by large faults, which must have rendered the extraction of masses of rock of practically any size an easy matter once the initial excavation had been carried out. The quarry is now full of great boulders, fallen, or brought down from the sides, and it seems likely that the blocks were partly dressed inside the quarry. At the top of the entrance passage there is ample evidence that the dressing was carried still farther after the block had been raised to desert level. From the quarry to the cultivation, wherever the desert track became difficult, embankments can be traced—often of very considerable size—some spanning gorges, others forming a flat track over the undulating desert.

Fig. 18. Holes cut for an unknown purpose in a quarry-face 15 feet above the present ground level. Ma'sara

Fig. 19. Alabaster quarry anciently known as Het-nub. The figure on the left of the entrance passage gives some idea of the height of the sides of the excavation, which in some parts is nearly 60 feet

The quarry inscriptions give the date at which stone was taken and the name of the person in charge of the work, but furnish very little information, if any, on the methods employed. One inscription at Tura, of the time of Amenophis III, tells how the king ordered galleries to be re-opened for the purpose of quarrying the fine limestone of 'Ayan after His Majesty had found that they had fallen into disrepair.[1] Another class of inscription, of late date, is frequently noticed in the Tura and Ma'sara quarries, consisting of a dedicatory note, usually in Demotic, to the deity who was supposed to preside over the gallery, and mentioning the name and affiliation of the works-foreman.[2] In these quarries, it seems that a record of the amount of stone extracted by a gang was kept on the roof, by marking on it the position of the vertical face at the beginning of the work.

At Gebelein, there is an interesting inscription of the time of King Nesibanebded, of the XXIst dynasty, in which it is stated that 3,000 men were sent there to get stone for repairing the canal-wall of Tuthmosis III, which had fallen into ruin.[3] There are suggestions that the king himself paid a visit to the work. In the *Het-Nub* alabaster quarry, there is an inscription of the time of King Mernerē', of the Vth dynasty, where a noble called Uni records that an offering-table of that material was cut and brought to the Nile in seventeen days, and that a large boat was prepared in the same short time to transport it.[4] In the tomb of Dhuthotpe at El-Bersha, it is stated that in the time of Senusret III, of the XIIth dynasty, a statue 13 cubits (22 feet) high—which would have weighed some 60 tons—was transported from the same quarry. The inscription comments on the difficulty of the road from the quarry to the river.[5] At Silsila, an inscription of Seti I records that the king, in his sixth regnal year, sent an expedition of 1,000 men to transport a monument, and that he paid the party 20 *deben* (4 lb.) of bread and two bundles of vegetables daily per head, and that each received two linen garments per month.[6]

The use of troops in the quarries was not entirely for keeping order among the quarrymen. In a letter of the VIth dynasty found at Saqqâra,[7] an army officer in charge of a detachment stationed at the Tura quarries, in reply to an order from headquarters to draw clothing, complains that when he recently spent six days with his men at the Residence-city, he

[1] BREASTED, *Ancient Records*, ii, § 875.
[2] SPIEGELBERG, *Annales du Service*, vi, pp. 219–33.
[3] BREASTED, *Ancient Records*, iv, § 629.
[4] *Ibid.* i, § 323; see also p. 34.

[5] *Ibid.* i, § 696; see also p. 85.
[6] *Ibid.* iii, § 206.
[7] For the translation and discussion of this letter see GUNN, *Annales du Service*, vol. xxv, pp. 242–55.

was not issued with clothing, and points out that to bring them over again for that purpose means the loss of a whole day's output of work. If the detachment had been merely on guard it is hardly likely that the officer would complain of a day in town. On the other hand, if he were responsible for so much stone being supplied in a given time, his complaint is more easily understood.

III

QUARRYING: HARD ROCKS

THE principal hard rocks used for building in ancient Egypt were the pink and grey granites from Aswân; the use of all the others is only occasional. Basalt was used in the temples of the Great and Second Pyramids of Gîza and in two of the Vth dynasty temples at Abusîr. The walls of the great cenotaph of Seti I at Abydos, commonly known as The Osireion, and the burial chamber in the pyramid of Amenemhēt III at Hawâra,[1] are of quartzite.

The schists from the Wady Hammamât, though much prized, and used for sarcophagi and statues, were never employed in building.

Granite, of various colours, occurs on the east bank of the Nile at Aswân, and the ancient quarries are found chiefly in the area bounded by the Nile and the Aswân-Shellâl railway. Boulders from which blocks have been detached by means of wedges can be seen by the thousand, and all over the area great embankments, which facilitated the transport of the blocks to the Nile, can be traced leading from the quarries.

Basalt is found at Abu Za'bal, near Khânqa, and there are outcrops at Post No. 3 on the Cairo-Suez road and also at Kerdâsa, near Gîza. It seems likely that the pavement of that material in the Great Pyramid temple came partly from Kerdâsa, though there is no definite proof of it.

Quartzite, the hardest of the Egyptian rocks, occurs at Gebel Ahmar (The Red Mountain) near Cairo, and there is another outcrop near Gebelein, from which the Colossi at Thebes are believed to have come.

Although quartzite is not often found used for building, its quarrying demands study, since the extreme hardness of some of its varieties necessitated a technique not used in the quarrying of the granites.

Two methods of quarrying were used on the granites, namely wedging and pounding with balls of dolerite. The earliest traces of wedges used for

---

[1] See PETRIE, *Kahun, Gurob, and Hawara,* p. 16, where he states that: 'the sepulchre is an elaborate and massive construction. The chamber itself is a monolith 267·5 inches long, 94·2 wide and 73·9 high to the top of the enormous block, with a course 18·5 high upon that, giving a total height inside of 92·4 ... The thickness of the upper course is 36 inches from its face, but the chamber itself is about 25 inches, according to the outside seen in the forced passage from the western well. It would accordingly weigh about 110 tons. The workmanship is excellent; the sides are flat and regular, and the inner corners so sharply wrought that—though I looked at them—I never suspected that there was not a joint there until I failed to find any joints in the sides. ...'

splitting granite, which can be definitely dated, are those on the back of
the roof-blocks in the pyramid of Menkewrē' at Gîza. The best place,
however, for the study of ancient wedging on the hard rocks is at Aswân,
though unfortunately none of the many examples can be dated. Here the
wedge-slots are oblong in shape, usually about three inches in length and
tapering sharply inwards, the interior being quite smooth. It seems that
these slots were not used with wetted wooden wedges (p. 18), since their
taper and the smoothness of their sides would cause the wood to jump out
rather than exert the required lateral pressure. The line of wedge-slots is

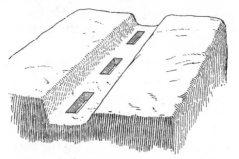

Fig. 20. Wedge-slots cut in a trench in the
granite quarries at Aswân.

frequently cut inside a channel (Fig. 20); the reason seems to be that the
surface of the boulders is more or less decomposed, and tends to crumble,
and would therefore give no hold to the wedges. Here and there one can
observe very large wedge-slots which may have been used for wooden
wedges made to expand by water. The cutting of wedge-slots almost inevitably involves the use of a metal tool, though a chisel-shaped piece of
dolerite, attached to a haft or held in the hand, might conceivably have
bruised them out. It must be borne in mind, however, that tool-marks are
found on certain granite quarry-faces at Aswân which could not have been
made with a stone. Certain scholars believe that the Egyptians could not
cut the hard rocks with a metal tool. Though this is surely erroneous, it is
certain that they could not cut the hard rocks with a chisel as they cut the
limestones, sandstones, and alabaster.

   In all ancient work on the hard rocks which has been left in an unfinished
condition, it is clear that the stone has been struck with a pointed tool,
and it is difficult to believe that this tool was of stone (Figs. 21 & 22). In
an unfinished schist statuette of Saïte date, in the Cairo Museum (Fig. 23),
the marks of the tool can be clearly seen; each blow has removed a small

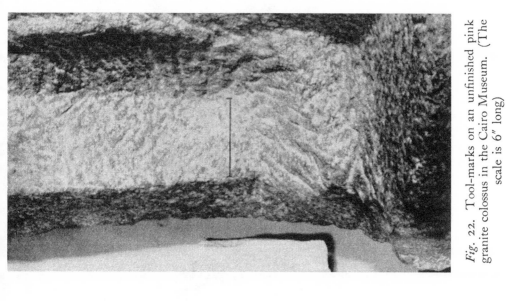

Fig. 22. Tool-marks on an unfinished pink granite colossus in the Cairo Museum. (The scale is 6″ long)

Fig. 21. Tool-marks on the 1st dynasty basalt stela of King Smerkhet in the Cairo Museum. (The scale is 6″ long)

Fig. 23. Unfinished schist statuette of Saïte date in the
Cairo Museum. (The scale is in inches)

Fig. 24. Modern quarrying of granite; Aswân. (Metre scale)

fragment of the stone without any apparent bruising, and a succession of a dozen blows or more can be traced without any evidence of wear on the tool.

Nowadays, the only metal used on the hard rocks is steel. To attempt to use copper, however much one may temper it by heating and hammering it, is to ask oneself whether the Egyptians did not use a tool of a hardness approaching tool-steel. It is tolerably certain that steel of this hardness was not known to the Egyptians. In the ancient language, the words for nearly all the metals have been identified, and it is incredible that the word for iron should have covered steel also. Further, cutlery such as razors, chisels, adzes, axes, &c., is almost always of copper.[1] Had hard steel been known, such implements would be expected to have been made from it. An examination by microscopic section of the copper of the ancient tools shows that they had never been raised to the annealing temperature, at which the crystalline structure disappears. Ancient copper tools are no longer of any great hardness; the edge of a chisel, for instance, soon burrs away if used to cut moderately hard limestone. If certain alloys, such as tin, are present, copper can be brought to a temper not far short of mild steel, but this is not sufficiently hard for cutting out the corners of, for example, a quartzite sarcophagus, which the Egyptians could do with great accuracy. It is possible that the Egyptians possessed the now lost art of giving copper a very high temper. Temper, which is a molecular or crystalline strain, can disappear in the course of time. We hear now and then of methods being discovered of giving copper a very high temper, but no definite details of any of them ever seem to come to hand.

The modern method of splitting granite by wedges gives us very little help in understanding the ancient technique. The tools used nowadays for cutting the hole to take the splitting medium in granite are called points, and are made from $\frac{5}{8}$-inch hexagonal steel bar. The bar is cut into lengths of 6 inches and drawn out by the smith to a blunt point, making it about 8 inches in length. The points are sharpened on a stone and are tempered after sharpening so that the hard skin shall not be ground away. Special precautions are taken by good workmen to harden as little as possible of the tool, so that, when it breaks off, a minimum amount of metal is lost. To ensure this, the points are cooled after heating by standing them in a stone trough made to hold water not more than an inch deep. The holes for the wedges are cut with these points, with the aid of a hammer of about 6 lb. in weight, and they are placed about 3 or 4 inches apart along the proposed

---

[1] A notable exception is the fine hunting-dagger of Tut'ankhamūn, which is of iron with a decorated gold handle.

line of fracture (Fig. 24). Each one is made about 2 inches deep, not oblong like the ancient examples, but of elliptical form, the longer axis being along the line of intended fracture. A good cutter will make his wedge-holes wider below the surface than at it, so that the point of the wedge or 'plug' shall be clear of the stone at the bottom. The steel plugs used to split the granite are from 3 to 5 inches long, and are of oval section, with a taper of three-quarters to half an inch along their length. They are placed in the holes with their larger diameter at right angles to the line of the fracture, and they are jammed tight by means of a hand hammer. They are then struck, in succession, one blow each with a sledge-hammer, very careful watch being kept for any sign that the crack is not running along the intended line. If this happens, it can often be corrected in time by starting a new series of holes leading back to the point at which the fracture began to go wrong. No traces of such corrections have been observed in the ancient work.

In modern work, care is taken that a block, once wedged off the parent boulder, falls conveniently outwards; but if this does not happen, it can always be freed by means of crowbars or tackle. In the Aswân quarries, blocks are sometimes seen which have been detached by wedges from the parent boulder, but have not fallen away from it. It seems that they were abandoned by the ancient quarrymen because they did not possess the metal crowbars which, apart from such implements as tongs actuated by tackle, would be the only means by which they could be conveniently removed. It was apparently easier to start work on another block than to cut large recesses in the stone into which the points of their wooden levers could be inserted for getting the block out.

Three forms of hammer used by the Egyptians are known, though no information is available as to which form was used to drive in the wedges. The first form is the ordinary sculptor's mallet (Fig. 264); the second is a club-shaped piece of wood, which was used both for chisels and for hammering in stakes (Fig. 36 & Fig. 61); and the third was a two-handled implement which was used for dealing heavy blows (Fig. 38), and of which actual examples (one in black granite) have been preserved.[1] It is likely that the last corresponded to our sledge-hammer. No hammer of modern type has been found, though it would be somewhat rash to assert that it was not known.

The second method of quarrying granite, used when the surface outcrop of boulders did not provide a block of suitable dimensions, is the process

[1] ENGELBACH, *The Problem of the Obelisks*, Fig. 10, p. 42.

which we have called pounding, the systematic bruising away of the rock
by means of balls of a very tough greenish stone called dolerite, which
occurs in some of the desert valleys between the Nile and the Red Sea.
These balls vary from 5 to 12 inches in diameter and weigh on the average
12 pounds. They were mostly used in the hand (Fig. 240, & Fig. 266),
though in making a great separating trench it is conceivable that they
were shod on to some form of haft and used much as a road-rammer
is used in Egypt to-day. It is known, at all events, that a dolerite pounding
tool was sometimes attached to a haft, since a piece of dolerite of the
XIth dynasty has actually been discovered, bound by leather thongs on to
two pieces of wood; it had been used in the excavation of the hill-tombs at
El-Deir el-Bahari [1] (Fig. 266). Since the process of pounding applied to
a large obelisk has been described by one of the writers in detail elsewhere,[2]
an outline of the method will suffice here.

To find a flawless piece of granite of any great size, it was necessary to
go down to a considerable depth. There are indications, at Aswân, that the
top layers of rock were removed by burning large fires on or against the
granite. These fires appear to have been banked up with crude brick.
When hot, granite can be made to break up by pouring water upon it,
which renders it so soft and crumbling that it can almost be broken away
with the fingers. The burning process, however, had to stop before it came
near to the block it was required to remove. The next step was to render
the top of the block more or less flat, which was done entirely by pounding
it with the dolerite balls. On the pyramidion of the great unfinished obelisk
at Aswân, traces of the use of these balls can be clearly seen (Fig. 25). Here
the surface is divided into squares of about twelve inches side. While this
was taking place—or so it appears—test-holes, of squarish form and about
a yard wide, were sunk at intervals along what was to be the separating
trench, probably with a view to finding out whether there were likely to
be any serious flaws in the block. This work may well have been carried
out by the best workmen working in short spells. These test-holes were
made entirely by pounding—a most laborious process in such a confined space.

When the work on the test-holes was well advanced, the separating trench
was begun. In the case of the Aswân obelisk this was nearly 300 feet long
in all (Fig. 26). The width of the trench is some 2 feet 6 inches, and vertical
lines, drawn in red ochre, can still be seen, which had been projected down

[1] From the excavations of the Metropolitan
Museum, New York. Now in the Cairo Museum.
[2] ENGELBACH, *The Problem of the Obelisks*.

from time to time, and which divide the trench into intervals of feet along its length. The appearance of a foot length of trench at the bottom was that of two circular depressions (Fig. 27). The method by which a maximum number of men might pound the rock without interfering with one another seems to have been that each man was made to work on a two-foot length of trench, the pounding being carried out in four positions, each man in the trench working in the same relative position in his portion, namely, pounding on the left and right half of his two-foot length of trench, both with his back to, and facing, the monument which was being extracted. Thus there would always be one foot between a man and his neighbours. When a man was pounding in one of the four positions he would have to remove, at short intervals, the granite powder or fragments which collected on the part on which he was working, and his natural procedure would be to brush the powder on to the portion of his work on which he was neither sitting nor pounding. A more efficient method could hardly be imagined.

The effect of such a system of work would be to leave the sides of the trench in the form of corrugations. The same appearance is also seen in a quarry-face above the Aswân obelisk (Fig. 28) from in front of which another monument of considerable size has been removed.

If, during any part of the work, a suspicious fissure or discoloration appeared, it was immediately given special attention, being pounded deeply along its length in order to determine whether it would be likely to prove a serious flaw as the work deepened. Parts of these grooves were often polished. Signs of this procedure can be seen in many places on the Aswân obelisk.

When the required depth in the separating trench was reached, the monument had to be detached from below. In the case of an obelisk, the use of wedges, whether hammered or wetted, would be certain to set up uneven strains along its length, and such a long thin block of stone would certainly not stand it. Hence it was also pounded out from below, presumably by driving galleries at intervals under it, filling them with packing of some kind, and then pounding out the remainder. This must have been the most laborious part of the whole work, since it would have to be done in a very cramped position. From a bed from which a smaller obelisk or similar monument has been removed, it seems that, though the two-foot job of each man was maintained, within that length the work was quite irregular, which is what would be expected. It is probable that monuments of the sarcophagus type were removed by wedging,

Fig. 25. The unfinished obelisk at Aswân. (From *The Problem of the Obelisks* (Fisher Unwin), *Fig.* 4)

Fig. 26. Top of the pyramidion of the unfinished obelisk at Aswân. (From *The Problem of the Obelisks* (Fisher Unwin), *Fig.* 5)

Fig. 27. Separating trench in the unfinished obelisk at Aswân.
(From *The Problem of the Obelisks* (Fisher Unwin), *Fig.* 11)

Fig. 28. The unfinished obelisk at Aswân, showing side of a separating-trench and a
quarry-face from before which another monument has been removed. (From *The*

but that the old quarrymen, having taken very great trouble in getting down to good rock and pounding out the separating trenches, preferred to take some more in pounding out the monument from below, if there were the slightest risk of it snapping across owing to the uneven strain which the wedges would be certain to set up.

The quarry-face above the Aswân obelisk (Fig. 29) mentioned on p. 28 shows many features of interest. Its surface is covered with lines and marks in red ochre made by the ancient quarrymen. Many of these marks

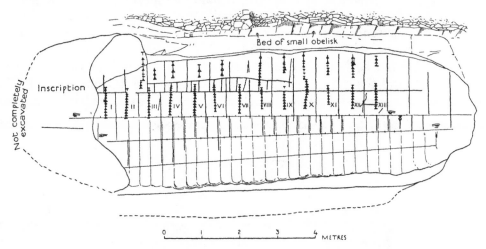

Fig. 29. Workmen's markings on the quarry-face near the Aswân obelisk.

are as yet inexplicable, but the vertical red lines clearly indicate that the corrugations, or foot-lengths, are connected in pairs. Further, it seems likely that the work on the separating trenches (of which this is, so to speak, the back wall) was measured by depth, and not by volume of stone extracted. The markings, which have the appearance of red chains, stop at about sixty-two inches above what was the bottom of the trench. From this it seems likely that, from time to time, the foreman stood a three-cubit rod [1] on the bottom of each section of trench and marked a small horizontal line in red ochre to indicate the position of the top of the rod. This line appears then to have been connected to those above it by an inverted Y. Within the squares (numbered in the drawing by Roman numerals) there are traces of inscriptions in red, now too faint to be legible. From evidence in other quarries, it is probable that these were the names of the gangs which

[1] This ingenious suggestion was made by Sir Flinders Petrie.

supplied the labour for that particular two-foot length of trench. At Gîza, on the roof-blocks in the relieving chambers in the Great Pyramid, and elsewhere where such names occur, there are generally fantastic phrases in praise of the king.

From experiments made by one of the writers at Aswân, the assertion of Queen Hatshepsowet on the base of her standing obelisk at Karnak that the quarrying of the two obelisks took seven months, is quite credible.[1]

Such is the process of pounding, which, from the presence of the dolerite balls, appears to have been used with certain modifications on all the hard rocks if a large monument were required. It is sometimes found used on the hard limestones, and in the quarry at Qâu (p. 18), the extraction of the monument, whatever it may have been, which lay above the large rectangular block (i.e. above AB, Fig. 15), was certainly carried out by pounding, and there is no trace of any metal tool having been used. The

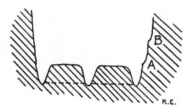

Fig. 30. Section through a separating-trench
in quartzite quarry at Gebel Ahmar.

rock tombs at Qâu were also largely hollowed out by pounding, the finer work only having been done by a metal tool.

In quarrying quartzite, the Egyptians seem to have used a metal tool in conjunction with the dolerite pounders, and an examination of the hardest red quartzite outcrop at Gebel Ahmar throws some interesting light on the methods employed. As was the case with granite, if a boulder could not be found from which a block could be removed by means of wedges alone, a separating trench was cut round the portion which it was required to remove, and the block was finally liberated by wedges, generally acting from below. The ancient wedge-marks still visible in the Gebel Ahmar quarries are not a series of slots, but form one continuous trench running nearly the whole length of the side of the block. It seems very likely that the wedges were of wetted wood (p. 18). Traces of these long wedge-trenches can be seen in many places at Gebel Ahmar;[2] they are about

[1] BREASTED, *Ancient Records*, ii, § 318.
[2] Wedge-trenches are occasionally met with in the

Aswân quarries, though we have no information as to their date.

Fig. 31. Quarry-face in the quartzite outcrop at Gebel Ahmar, showing the remainder of a separating-trench by means of which a block has been extracted from above ABCD. (The scale is 8 inches long)

Fig. 32. Front view of a quarry-face at Gebel Ahmar, showing the ridges which are peculiar to the quarrying of quartzite. (The scale on the photograph is one foot)

Fig. 33. Vertical quarry-face at Gebel Ahmar, showing marks made by a pointed tool. (The scale is in inches)

three inches deep with a slight taper. Their sides are fairly smooth, but not polished.

The method of cutting a separating trench in quartzite shows certain differences from that used for granite. In granite, balls of dolerite can be used to bruise away the stone, and, if necessary, a shaft can be sunk in granite by this means. The hard red quartzite of Gebel Ahmar cannot, it appears, be broken up to any great extent by the pounders alone, though a projecting lump can easily be jarred off by a heavy blow. The method of making a separating trench in quartzite seems to have been as follows: a line of holes, as close together as possible and about two inches in depth, was made with a blunt-pointed tool along what were to be the two walls of the trench, and another line of holes was made midway between them (Fig. 30). These lines of holes became, as it were, miniature separating trenches, leaving two ridges of rock between them, which were then jarred off by blows with the dolerite pounders, and the process repeated. On the quarry-face of the greater part of this outcrop there is a slope of some 10° from the vertical (Fig. 31). If a tool similar to a mason's pick were used, it will be seen (Fig. 30) that the tendency would be, after the ridges had been jarred off, to make the next line of holes slightly inside the previous series owing to the width of the tool, which would have the result of leaving a ridge along the side of the separating trench (Fig. 31, *A,B*). These ridges are a feature of the Gebel Ahmar quarries (Fig. 32) and are not, to the writers' knowledge, met with anywhere else. In a few places, the walls of the separating trenches were maintained vertical, which must have been the case for all large work. In these cases no ridges are left on the quarry-face (Fig. 33).

An examination of the hard rock quarries shows clearly that a pointed tool, most probably of metal, was used in cutting them, and it is likely that it was some form of mason's pick. The alternative would be a point used in conjunction with a mallet. It is a well-known fact among stonemasons that moderate blows with a heavy tool have a greater effect than the most violent blows with a hammer on such a light tool as a punch, though the reason is somewhat hard to explain.[1] With what success, however, the hard rocks could be cut with a copper pick is a matter of speculation, and might well be tried by a person accustomed to using this tool, since skill plays a very

[1] An analogous case is seen in the manner in which a quarryman uses his heavy crowbar or 'jumping iron' when pounding out a long cylindrical bore to take a blasting charge. He brings it down fairly gently, and checks the rebound with his hand, giving the tool a slight turn before delivering the next blow. However hard it is jabbed down, in the hands of an amateur hardly any progress is made, and the edge of the tool is quickly rendered useless.

great part in such work, and a practised workman might obtain results which could not be obtained by an amateur. Another indication that a mason's pick may have been used on the hard rocks is found in certain statues, where deep and narrow recesses had to be cut, such as those into which it was intended to inlay the eyes, into which such a tool could not enter. If a point struck by a mallet had been used to dress the statues, it would have been equally easy to cut out any form of recess with it. An examination, however, shows that they were always cut out by means of a series of tubular drills of different sizes (Fig. 245). This shows that the temper of their metal was such that it was found quicker, or more economical, to drill than to cut the hard rocks. It is very likely that the point was used, but as sparingly as possible.[1]

At Gebel Ahmar there is a ledge of rock on which a quarryman has given a few taps with his tool, either idly or to try its temper or point. It seems that the point of the tool he used was about one-eighth of an inch square. It will be noticed that the loss of metal in a blunt-pointed pick would not be great, though one would expect that very frequent repointing would be necessary if the metal were copper as we know it. The marks on an unfinished schist statuette (Fig. 23), however, show that the tool retained its point for a remarkably long time. The Egyptian stoneworkers never seem to have used a metal tool on the hard rocks if pounding were in any way possible, unless it was to cut the holes for the use of wedges. Granite false-doors and other monuments are often seen with the internal angles left slightly rounded. In the tomb of Rakhmirē' there is a scene in which a craftsman is shown pounding out the details of a uraeus on the head of a sphinx (Fig. 241) with what appears to be a stone chisel which he holds in his hand. In the great pink granite colossus, lying unfinished in the Aswân quarries, it can clearly be seen that it had been dressed to what were to be almost its final contours by the use of pounders alone.

A few records exist concerning the numbers of men sent out on expeditions for quarrying the hard rocks. In the IX–Xth dynasties, a little-known king called Imhōtep[2] sent his son Kenūfer to the Hammamât quarries with 1,000 men of the palace, 100 quarrymen, 1,200 soldiers, 50 oxen, and 200 asses. King Menthuhotpe IV,[3] of the XIth dynasty, also sent an expedition there, numbering 10,000 men, to quarry stone for a large sarcophagus. Here, 3,000 sailors, probably pressed from the Delta provinces, were used to move the lid, which measured 13 feet 10 inches by 6 feet 5

---

[1] ENGELBACH, 'Evidence for the use of a mason's   [2] BREASTED, *Ancient Records*, i, § 390.
pick in ancient Egypt', *Annales du Service*, xxix.   [3] *Ibid.* i, § 448.

inches by 3 feet 2 inches, from the quarry to the river. It is related that 'not a man perished, not a trooper was missing, not an ass died and not a workman was enfeebled'.

In the reign of King Amenemhēt III,[1] of the XIIth dynasty, an official of the same name was sent on the Hammamât expedition for 10 statues, each 8 feet 8 inches high. The personnel was as follows:

Necropolis soldiers	20
Sailors	30
Quarrymen	30
Troops	2,000

Under Ramesses IV,[2] there is a record of a very large expedition of 8,362 persons sent to the Hammamât quarries for monumental stone. The personnel consisted of the following:

High Priest of Amūn, Ramesse-nakht, Director of Works	1
Civil and military officers of rank	9
Subordinate officers	362
Trained artificers and artists	10
Quarrymen and stonecutters	130
Gendarmes	50
Slaves	2,000
Infantry	5,000
Men from 'Ayan	800
Dead (excluded from total)	900
	8,362

Before leaving the subject of quarrying, it may be remarked that the Egyptians classified their rocks by appearance, hardness and locality, describing the material of a temple, for example, as fine white limestone from 'Ayan. They had a word for limestone, for sandstone, for alabaster and for granite, but the basalts and the schists seem to have been all grouped together under one term. On the other hand, each variety of quartzite from Gebel Ahmar had its separate name.

---

[1] BREASTED, Ancient Records, i, § 710.   [2] Ibid. iv, § 466.

IV

TRANSPORT BARGES

THE working of hard stones, such as granite, diorite, basalt, and quartzite, was practised by the ancient Egyptians from predynastic times, and it can be assumed that the knowledge of building barges to carry heavy weights kept pace with the demand for blocks of ever-increasing size, since the quarries for these materials are, generally speaking, far away from the large towns and cemeteries where they were wanted, and the Nile was the sole practical means of communication. As far back as the XIIth dynasty we know of a block of quartzite being transported, presumably from the Gebel Ahmar, near Cairo, to the entrance to the Fayyûm. This was the monolithic chamber, weighing over 100 tons, which King Amen-emhēt III built into his brick pyramid at Hawâra (p. 23). In the New Kingdom ships were built which carried blocks of granite, some of which weighed nearly 1,000 tons, from Aswân to Luxor. The determination of the nature of these 'august barges', as the Egyptians called them, is one of the most perplexing problems that the archaeologist has to face.

Knowledge of ancient Egyptian boats is derived from the following sources: (*a*) the overall dimensions given by Egyptian and classical writers, (*b*) the actual royal barges of the XIIth dynasty found buried near the north pyramid at Dahshûr, (*c*) the many representations of ships, both pleasure and cargo, mostly of modest dimensions, on the walls of ancient tombs, and (*d*) the numerous models of pleasure, cargo, and religious boats which, in certain dynasties, were deposited in the tombs.

The only dimensions of ancient boats which have come down from purely Egyptian sources are those of the cedar-wood boat of King Sneferu,[1] which was of 100 cubits length (172 feet); that of Uni,[2] of the VIth dynasty, which was of acacia-wood and measured 60 cubits (103 feet) in length with a beam of 30 cubits (51 feet 6 inches), which he, in his autobiography, asserts took only seventeen days to construct, and that made for King Tuthmosis I,[3] which measured 120 cubits (206 feet) in length with a 40-cubit (69 foot) beam. Classical writers give certain indications of the sizes of ancient ships and boats, but these afford very little real help. Pliny, in

---

[1] SCHÄFER, *Ein Bruchstück altägyptischer Annalen*, p. 30.
[2] BREASTED, *Ancient Records*, i, § 323.
[3] *Ibid.* ii, § 105.

Fig. 34. Royal barge of the XIIth dynasty from Dahshûr, constructed of small pieces of wood tenoned and mortised together. Length 33 ft. 5 in.

his Natural History (Bk. xxxvi, cap. 14) relates that King Ptolemy Philadelphus constructed a boat to carry an obelisk to Alexandria, and describes
how a canal was cut passing under the obelisk (which was lying on its side)
and how the boat was unballasted beneath the obelisk and so took its weight.

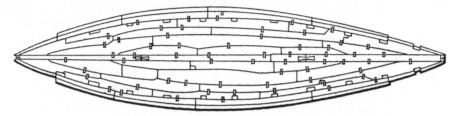

Fig. 35. Plan of royal barge shown in Fig. 34. (After Reisner.)

Fig. 36. Boat-builders, from the XIIth-dynasty tomb of Khnemhotpe at Beni Hasan.

Other classical writers state that great ships were built for use on the Nile,
and vessels of 40 and even 50 banks of oars are mentioned; but if by the
word 'banks'[1] these writers mean ranges of rowers, one above the other,
their statements can hardly be believed. Ptolemy Philadelphus is said to
have caused a boat to be built over 100 yards long and nearly 10 yards
high, and it is also said that the number of ships belonging to this king
exceeded those of any other king, and that he had two of 30 banks and
four of 14 banks.

[1] Torr, *Ancient Ships*, pp. 3–9 and 15.

Not only could the Egyptians build boats to carry immense weights, but they could also build sea-going ships.[1] There are records of such ships as far back as the time of King Sneferu of the IVth dynasty, and, throughout the history of Egypt, expeditions were constantly being sent by water to the coasts of Palestine and round the eastern Mediterranean. In the reign of Queen Hatshepsowet a large trading expedition was dispatched to the Red Sea coast (*Punt*): a magnificent series of sculptures of the ships can be seen on the walls of her temple at El-Deir el-Bahari.[2]

The Egyptians seem to have constructed their boats in a manner quite different from those of modern times, and their methods were the subject of comment by Herodotus (II, 96). The skin of the boat was made by jointing small pieces of wood together, the boat itself being entirely without ribs. Lateral rigidity depended on the thwarts, which ran from side to side

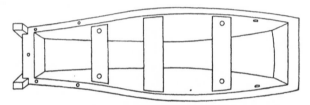

Fig. 37. Model punt-like boat. Middle Kingdom. From El-Bersha. The mast is shown in Fig. 43.

throughout the length of the boat. From the sculptures it can be seen that longitudinal rigidity was obtained by a rope or ropes attached at each end, which passed over two stiff stays at one-third and two-thirds of the way along the boat, thus forming what the English call a 'queen-truss' and the Americans a 'hog-frame'. A royal barge from Dahshûr, of the XIIth dynasty, which is preserved in the Cairo Museum (Fig. 34) is thus constructed, the pieces of wood being tenoned and mortised into one another (Fig. 35),[3] and the construction of such boats is frequently depicted in the tomb scenes (Figs. 36[4] & 38), the patchwork principle by which they were built up being sometimes very clearly shown.

Whether the construction of boats from small pieces of wood was the result of the lack, in Egypt, of trees which would provide long planks,[5] is

[1] See Breasted, *Ancient Times*, Fig. 41, p. 58.
[2] Naville, *The Temple of Deir El Bahari*, vol. iii, Pls. LXXII et seq.
[3] Reisner, *Models of Ships and Boats (Catalogue Général du Musée du Caire)*, p. 84.
[4] From Lepsius, *Denkmäler*, ii, Pl. 126.

[5] The home-grown wood used by the ancient Egyptians seems to have been mainly the *Acacia Nilotica* (Arab. *ṣunṭ*) and the sycamore-fig (Arab. *gammêz*). Neither of these is suitable for providing long planks.

Fig. 38. Boat-building in the Vth dynasty, showing the tools used by the carpenters. Saqqâra

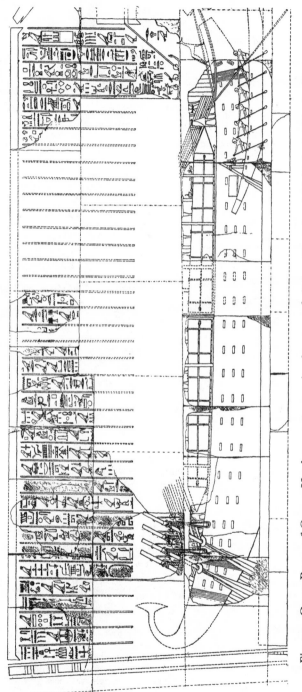

Fig. 39. Great Barge of Queen Hatshepsowet, carrying her two obelisks. Temple of El-Deir el-Bahari, Thebes.

uncertain, but we have here a very important question, and one very difficult of solution in the light of our present knowledge, namely, were all
ships built on the patchwork principle or not? Here, model boats afford
little help, since they are either constructed solid or are of punt-like form
(Fig. 38[1]), and obviously represent quite small craft. It seems likely that
ships of moderate dimensions were so constructed.

   The only contemporary representation of a great weight-carrying barge

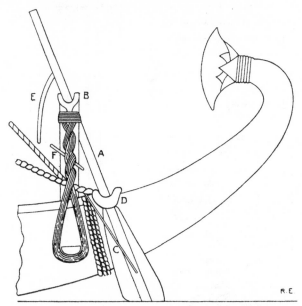

Fig. 40. Steering-system of the trading ships of Queen
Hatshepsowet. XVIIIth dynasty. Temple of El-Deir
el-Bahari.

is at El-Deir el-Bahari, where a vessel is represented carrying two obelisks,[2]
placed end to end upon it (Fig. 39). It will be noticed in the illustration
that in spite of its great size the traditional shape of Egyptian boats is maintained. This was almost certainly derived from the primitive papyrus
boats which are still used on the upper Nile (Fig. 41). The vessel is also
represented with three ranges of what appear to be thwarts. In attempting
to glean information from this sculpture several points must be appreciated;

[1] Middle Kingdom, from El-Bersha.
[2] It used to be considered that the two obelisks were
not those of Karnak, but two which once stood
before the Temple of El-Deir el-Bahari. Recent
excavations, however, have tended to show that
this temple was never provided with obelisks.

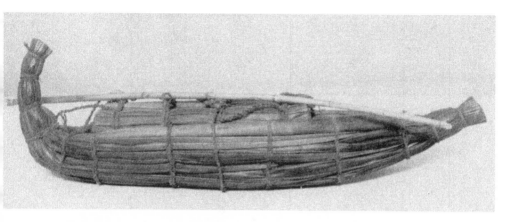

Fig. 41. Model of a papyrus boat (modern) used at the present day on the Upper Nile. It is likely that the traditional form of Egyptian boat was derived from this form of craft

Fig. 42. Steering-paddle of a model ship of the XIth dynasty from the tomb of Meketrē' at Thebes. Now in the Cairo Museum

first, it must be remembered that the artist cannot have done his drawing on the walls from the actual ship, but probably did it from rough sketches which he had made when he attended the arrival of the obelisks in his official capacity. It does not follow that he must have been an expert in the details of ships; indeed, from his representation of the tackle, it seems he was not. Secondly, it must be borne in mind that an Egyptian, when he wished to represent one object in another, or behind another, often drew what was inside or further away above that which was nearer. The drawing

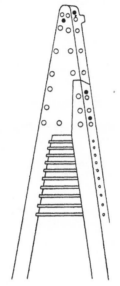

Fig. 43. Mast-head of the small punt shown in Fig. 38. Middle Kingdom; El-Bersha.

of the barge gives an impression of top-heaviness with the two obelisks standing high on its deck, and it may be that they were really inside. This rather rules out the possibility that it was provided with three ranges of thwarts, unless the barge was built round the obelisks. While waiting for further information the student is completely in the dark regarding the internal structure of these barges; the patchwork method of boat-building seems hopelessly inadequate to resist the strain that the skin of the barge would have to endure, even if it were internally stiffened with a series of queen-trusses. It is extremely likely that the great barges were solid rafts made of tree-trunks, the whole raft being, if necessary, shaped to give it the appearance of a true ship. Their draught would admittedly be great, but not too great to prevent them passing down the Nile during flood-time. The time given by Uni (p. 34)—seventeen days—for the construction of his great barge makes this more likely, since it is not stated that it was shaped like those of Queen Hatshepsowet.

The presence of the series of stiffening-ropes on the great barge of Queen Hatshepsowet and the thwarts, the ends of which can be seen in the sculpture (Fig. 39), need not necessarily imply that the boat was of the patchwork kind already described. In a solid raft, made of many logs lashed together or otherwise attached, they would probably have been very

Fig. 44. Lowering the mast of a IVth-dynasty ship. Tomb of Abibi. From Saqqâra; now in the Cairo Museum.

necessary for keeping it in shape. Such is the unsatisfactory state of our knowledge of the great weight-carrying barges.

The peculiar form of Egyptian ships necessitated a steering-system different from that used to-day. In moderate-sized craft the primitive method was often used, of one or more paddles held in the hand and passing through rings or lashings at one side of the boat as far aft as convenient. Such a 'rudder' can be seen in the tomb of Seknūfer of the IVth dynasty at Gîza.[1] Here three men are steering with three paddles. A development of the primitive method was to attach the steering paddle at two points and to fit it with a lever, corresponding to a tiller, by which it could be rotated, the paddle thus acting as a true rudder. Sometimes one paddle only was used; at other times there was one on each side. In the great

---

[1] Lepsius, *Denkmäler*, ii, Pl. 28.

obelisk-barges of Queen Hatshepsowet there are two fixed paddles on each side. It must not be supposed that, in the two-paddle steering-gear, the boat was made to turn, say, to the right, by turning the right paddle-blade flat to the water and vice versa; to do this would be to unship the paddle. Each paddle was a separate steering rudder, and the boat could be steered both to port and starboard by either of the paddles, though the efficiency of each differed according to which direction the turn was made. In the

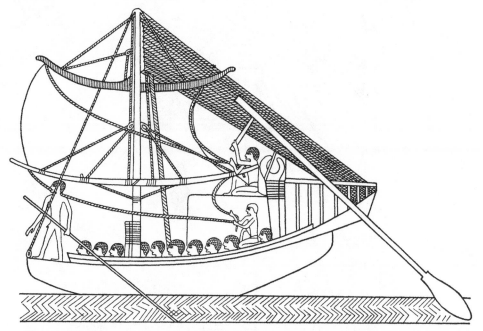

Fig. 45. IVth-dynasty ship under full sail. Tomb of Ipi. From Saqqâra, now in the Cairo Museum.

multiple-paddle steering-system, one paddle formed the rudder and the others reinforced it when necessary. Fig. 40 shows the details of a fixed rudder in one of the ships which took part in the expedition to Punt under Queen Hatshepsowet. The paddle (*A*) rests in an upright (*B*), and is prevented from slipping down by a stay-rope (*C*), and held to the hull by a loop (*D*). The paddle was rotated on its axis by a tiller (*E*). Ropes attached to one side of the boat passed over the steering-pillars and thence down to the other side, being tightened by a tourniquet (*F*). The last was to ensure rigidity in the steering-gear.

When the primitive form of Egyptian boat was departed from, and the stern made rounded, one steering paddle only was required, placed centrally. A steering-system of this type is seen in a model ship of the XIth dynasty from the tomb of Meketrē' at Thebes (Fig. 42).[1] The object of the attachment at the top of the supporting-pillar was to prevent the rope

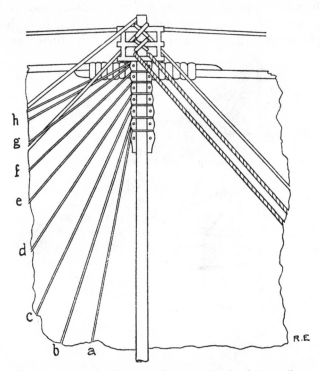

Fig. 46. Composite drawing of the mastheads of the trading-ships of Queen Hatshepsowet with sails hoisted. XVIIIth dynasty; El-Deir el-Bahari.

holding the paddle from slipping down the pillar. A single steering-paddle with the traditional form of boat is sometimes seen, as, for example, in the XIIth dynasty tomb of Khnemhotpe at Beni Hasan. In this case the supporting pillar is very high, and the long paddle is attached almost at the end of the pointed stern by a loop of rope.[2]

In the Old and Middle Kingdoms boats seem usually to have been fitted with double masts, which were joined at the top by transverse metal or

[1] Now in the Cairo Museum. See *Bulletin of the Metropolitan Museum of New York*, 1918–20. [2] LEPSIUS, *Denkmäler*, ii, Pl. 127.

wooden rods (Fig. 43 [1]). The earliest known boat of this type is found in a sculpture of the time of King Sneferu of the IVth dynasty. In the tomb of a noble called Abibi, also of the IVth dynasty, there is a scene[2] in which men are in the act of lowering the double mast (Fig. 44), and what is pro-

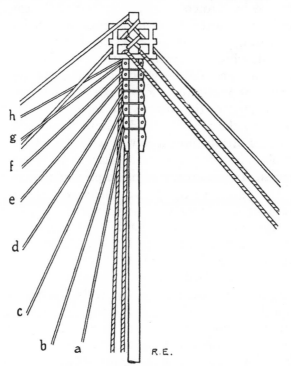

Fig. 47. Composite drawing of the mastheads of Queen
Hatshepsowet's trading-ships with sails lowered.

bably a double-masted ship under sail is found in the tomb-sculpture of a IVth dynasty noble called Ipi[3] (Fig. 45).

From other scenes it appears that, when a mast was lowered, it was stacked, together with the yards, on two upright wooden supports, whose tops were recessed in the same manner as those which supported the steering-paddles. It seems likely that the cross-members of the double-masted ships, seen at the mast-head, were used as attachments for the numerous stays which passed from the mast to the stern.

A study of the means by which the sail was hoisted is of great importance,

[1] From Reisner, *Models of Ships and Boats* (*Catalogue Général du Musée du Caire*), p. 54.
[2] Now in the Cairo Museum (*Cat. Gén.* No. 1419).
[3] Now in the Cairo Museum (*Cat. Gén.* No. 1536).

since it is the only direct evidence concerning the lifting-tackle used by the Egyptians, and has to be very carefully taken into account in the inquiry into the methods used in masonry for laying the blocks.

In the ship of Ipi (Fig. 45) the tackle is fairly clearly indicated. The mast is stayed by numerous guys attached to the stern and one to the bow. (In the single-masted boats there must surely have been stays on either side, though they are never represented in the sculptures nor in the models.) The sail, which is provided with yards above and below, is controlled by two men, one holding two lines attached to the top yard and the other a similar pair attached to the lower yard. The lower yard is supported by four stays attached to the mast. The top yard, and consequently the sail, is supported by the main halliards, though in this scene neither their number nor means of attachment is very clear. In this ship there are no pulleys; their place is taken by what appear to be wooden or metal rings lashed to the mast. In the great trading ships of the XVIIIth dynasty from El-Deir el-Bahari (Figs. 46 & 47[1]) it will be seen that only two halliards are used to support the upper yard when under sail, the rest (a–h) being merely stays supporting the lower yard. The curious attachment on the mast to which they are made fast cannot therefore be a series of pulleys, as some have supposed. At the only place where a pulley would be expected none exists, the main halliards merely passing over what may be a smooth metal frame lashed to the peak of the mast. In a ship of any size it is clear that the friction between the halliard and the member over which it passes would be very great when the sail was being hoisted. Since it is known, from other scenes, that men were accustomed to stand on the lower yard, it is not unlikely that they helped the hoisting of the yard by pushing from below, the halliard taking up the slack rather than doing much actual pulling. This may well be the reason for the apparently excessive number of stays supporting the lower yard, which is remarkable in all the models and sculptures of sailing-ships.

Although hundreds of models and pictures of sailing boats are known, a pulley occurs in none of them, at any rate in dynastic times,[2] and the evidence brought forward suggests that pulleys were unknown. Further, if they had been used, in building, for lifting the blocks of stone, it would

---

[1] In these drawings the corresponding stays on the right side of the mast have been omitted for the sake of clearness. In the sculptures of Hatshepsowet's boats the stays which held up the upper yard when the sail was lowered are mostly left out, only a few being shown hanging loose when the sail was hoisted. It must also be remembered that the fittings of the mast are turned by the artist through a right-angle so that the sail may be visible.

[2] The earliest known pulleys in Egypt are of Coptic or Roman date.

be expected that a model pulley would have been found in the foundation deposits under the temple walls among the tools, rollers, baskets, brickmoulds, and other objects which were so frequently placed there; yet none is found.

The examination of Egyptian boats in greater detail is outside the range of this volume. The reader who wishes to study all classes of Egyptian craft should consult REISNER, *Catalogue Général du Musée du Caire; Models of Ships and Boats*, and BOREAUX, *Étude de Nautique Égyptienne* (*Mémoires de l'Institut français*). The latter has recently appeared (1927), and gives the references to all articles on Egyptian shipping which have been published.

V

PREPARATIONS BEFORE BUILDING

IN ancient times, as at the present day, many things had to be done before the mason could begin building. Plans—perhaps models—of the proposed building had to be submitted to the king, who, either personally or by deputy, formally set out the limits of the building, conducted the foundation ceremonies, and made the necessary sacrifices to the god to whom the building was to be dedicated. The architect, in his turn, after preparing the plans, had to organize a constant and sufficient supply of stone from the quarries, which were often far away, and after the preliminary formalities were over accurately set out the lines of the proposed walls.

It is fortunate that a considerable amount is known about the preparations made before building. Actual plans and models have been preserved; temple sculptures give some idea of the nature of the foundation ceremonies, and tomb-scenes occasionally give glimpses of the ancient methods of measuring land. For the rest, the student has to rely on deduction based on observed facts.

It seems certain that there were palace archives where plans of temples were preserved, since in one of the crypts at Dendera an inscription states that the plan of the temple was found, written in ancient characters, in the palace of King Pepi. Another passage relates that a restoration had been made by King Tuthmosis III after a plan had been found dating to the time of King Khufu. The ancient references of this kind are always rather vague, and make the reader wonder whether he is reading facts or whether it is entirely an invention of the priests in order to magnify their office and all things connected with it.

The Egyptians were able to draw an object from different aspects,[1] showing side- and end-elevations, for example, but only one drawing has been preserved as a definite proof of this (Fig. 48).[2] It seems to have been found

---

[1] It is doubtful how far the Egyptians made use of sectional drawings in the construction of their buildings. A truly sectional representation of a house, showing the contents of each storey, is known in the New Kingdom (MACKAY, *Ancient Egypt* (1915), p. 171). It is by no means certain that the scene in the tomb of Rakhmirēʿ (Fig. 86, p. 92), which ap-

pears to represent a sloping embankment leading up to the top of three columns embedded in brickwork, is meant to be sectional; it is quite as likely that the embankment is meant to be in elevation and the remainder in plan.

[2] PETRIE, *Ancient Egypt* (1926), p. 24. Now at University College, London.

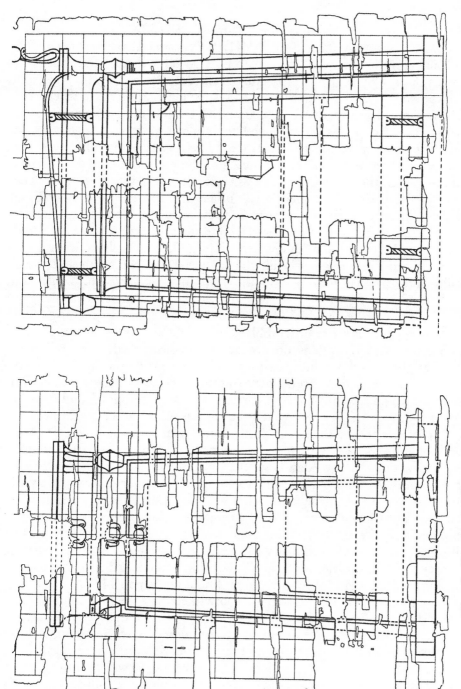

Fig. 48. Front and side elevations of a shrine on papyrus. XVIIIth dynasty. From Ghoráb.

at Ghoráb, and may date to the XVIIIth dynasty. It represents the front
and side elevations of what appears to be a portable shrine, and is drawn in
black ink on a piece of papyrus squared in red. Plans of tombs and estates
are known, often showing the doorways, pylons, altars, &c., in elevation
on the same drawing, somewhat after the fashion of medieval European
maps. The Egyptians dimensioned their plans more or less, but, to our
eyes, the dimensions are very meagre. Their style of building was more
traditional than that of modern times, and the more traditional the type of
building was, the less was the necessity for entering into a mass of detail.
Most of the plans of the ancient Egyptians are, however, amply sufficient
to give a clear understanding of what they were intended to represent.

Though the use of a squared surface for an architectural drawing is only
known in the example already cited (Fig. 48), it must have been a very
usual procedure, since nearly all the sculptures on the tomb and temple
walls were originally drawn by the artist over squares (p. 199). The
method of drawing the squares was to measure equal increments along the
edges of the surface on which the work was to be done by touching it with
a string, dipped in ochre or lampblack, and held at corresponding points at
each side. The Egyptians do not appear to have made very much use
of a straight-edge for ruling; neither the reed-pens nor, in most cases, the
surfaces on which they drew, were suitable for the ruling of a line. The
cords by which the squaring was carried out are sometimes found wrapped
round the reed-brushes with which the colours were put on (Fig. 266).

One of the most interesting dimensioned plan-elevations is that of the
tomb of Ramesses IV, on a papyrus preserved in Turin (Fig. 49), which is
very fully discussed by Alan Gardiner and Howard Carter in the *Journal
of Egyptian Archaeology*, vol. iv. In the plan the doorways are shown in
elevation, the openings being painted yellow. Inscriptions in hieratic give
the names of the various parts of the tomb and their dimensions in cubits,
palms and digits (p. 63). The plan measures 33·8 by 9·6 inches. The
hillside in which the tomb was cut is symbolized by a brownish surface
covered with a multitude of bead-like dots in rows of red and white alternately, recalling our ideas of *hatching*. Though it is possible that the corresponding under-side was similar, it is more likely that the line bounding the
dots below was horizontal, and represented the bottom of the mountain,
which, like the doors, was meant to be considered as an elevation, though
the tomb was in plan. The plan of the tomb itself is, to our modern ideas,
rather unsystematically dimensioned; for instance, it will be noticed that
the thickness of the doorways is not given. Walls were indicated by two

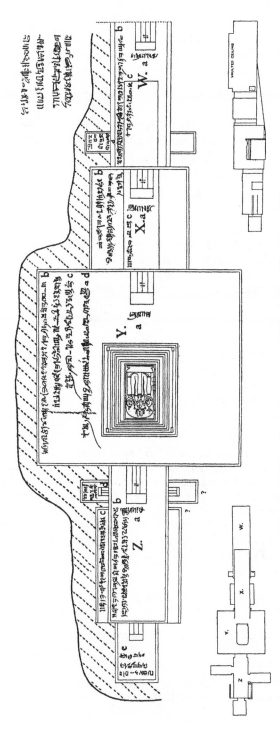

Fig. 49. Ancient plan, on papyrus, of the tomb of Ramesses IV.

parallel lines a short distance apart. A very interesting part of the tomb is Chamber Y, where, round the sarcophagus, are five series of walls which cannot be anything but a 'nest' of shrines similar to those found intact in the tomb of Tut'ankhamūn. It will be seen that, immediately within the outer shrine, there is something supported by four corner-posts, which is bounded by a single line, and not by the two lines which serve to indicate the wall; this is almost certainly the pall covering all the inner shrines. It may be of interest to give the translations (after Dr. Gardiner) of the hieratic notes on the drawing.

CHAMBER W:

   a. *Its door is fastened.*[1]
   b. *The fourth (corridor) of length* 25 *cubits; breadth of* 6 *cubits; height of* 9 *cubits* 4 *palms; being drawn in outlines, graven with the chisel, filled with colours and completed.*
   c. *The slide of* 20 *cubits; breadth* 5 *cubits* 1 *palm.*
   d. *This chamber is of* 2 *cubits; breadth of* 1 *cubit* 2 *palms; depth of* 1 *cubit* 2 *palms.*

CHAMBER X:

   a. *Its door is fastened.*
   b. *The Hall of Waiting of (length)* 9 *cubits; breadth of* 8 *cubits; height of* 8 *cubits, being drawn in outlines, graven with a chisel, filled with colours and completed.*
   c. *End of the sarcophagus-slide of three cubits.* (This dimension is misplaced, so as to make room for the elevation of the door.)

CHAMBER Y:

   a. *Its door is fastened.*
   b. *The House of Gold wherein ONE*[2] *rests, of* 16 *cubits; breadth of* 16 *cubits; height of* 10 *cubits, being drawn in outlines, graven with the chisel, filled with colours and completed, being provided with the equipment of His Majesty on every side of it, together with the Divine Ennead which is in the Dēʿet* (Underworld).
   c. *Total, beginning from the first corridor to the House of Gold,* 136 *cubits* 2 *palms.*
   d. *Beginning from the House of Gold to the Treasury of the Innermost,* 24 *cubits* 3 *palms. Total,* 160 *cubits* 5 *palms.*

CHAMBER Z:

   a. *Its door is fastened.*
   b. *The Corridor of the Shabti-place of* 14 *cubits* 3 *palms; breadth of* 5 *cubits; height of* 6 *cubits* 3 *palms* 2 *digits, being drawn in outlines, graven with the chisel, filled with colours and completed. That south of it as well.*
   c. *The Resting Place of the Gods, of* 4 *cubits* 4 *palms; height of* 1 *cubit* 5 *palms; depth of* 1 *cubit* 3 *palms* 2 *digits.*

     [1] Probably meaning 'Doorway to be fitted with door and bolts'.
     [2] A common term for the King.

Fig. 50. Plan, on limestone, of what is probably the tomb of Ramesses IX, from the Valley of the Kings at Thebes. (Cairo Museum)

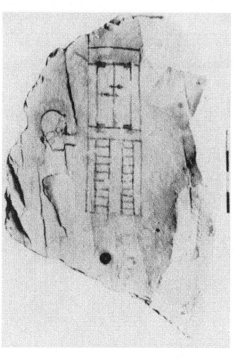

Fig. 51. Plan of a building on a limestone flake from the Valley of the Kings at Thebes. New Kingdom

Fig. 52. Plan-elevation of the door of a shrine approached by a double flight of stairs, on a limestone flake from the Valley of the Kings at Thebes. New Kingdom

d. *The Left Hand Treasury of* 10 *cubits; breadth of* 3 *cubits; height of* 3 *cubits* 3 *palms.*

e. *The Treasury of the Innermost of* 10 *cubits; breadth of* 3 *cubits* 3 *palms; height of* 4 *cubits.*

The measurements, as given on the papyrus, agree fairly well with those actually observed in the tomb of Ramesses IV, some being exact, while in other cases it is more probable that alterations in the original scheme are the reason for the observed differences than that they are errors. In the illustration the actual plan and section of the tomb of Ramesses IV are given for comparison with the ancient plan. On the reverse side of the papyrus there is a series of dimensions also referring to a royal tomb, but those that can be compared with the plan of the tomb of Ramesses IV are totally at variance with it. It is likely that they refer to a different tomb altogether.

Another plan of a royal tomb, almost certainly that of King Ramesses IX at Thebes, was found on a piece of limestone in the Valley of the Kings (Fig. 50). It measures 32·7 inches in length. The drawing is in red, the space between the parallel lines being filled in with white. The jambs and lintels of the doors are painted yellow. This plan should not be looked upon as the architect's original plan of the tomb, but rather as a sketch-plan for the guidance of the workmen.

In the same category must be placed two limestone ostraka also from the Valley of the Kings at Thebes and now in the Cairo Museum. The first (Fig. 51), drawn in red ink, represents a building whose roof was to have been supported by four columns of rectangular section. The door is shown laid out flat in the usual Egyptian manner at the top of the plan, but without any details. At the bottom of the plan is a note in hieratic which reads: 'Breadth, 15 cubits'. Opposite the top right-hand column is the figure '8', which probably formed part of a similar note reading 'Length, 18 cubits'. It seems likely that the artist intended to have a space of four cubits between the walls and the columns and between the columns themselves, and that, when he totalled up the number of cubits in the length of the building by making small 'ticks' on the right-hand wall, he found that it came to 16 cubits and not 18, and that he thereupon erased his note on the length. The proportions of the section of the columns as drawn on the ostrakon are about 2 to 1½. Assuming that these were the desired dimensions, and that the same spacing was to be observed in the breadth as in the length, the former amounts to 15 cubits as stated by the artist. Traces of 'ticks' can also be observed on the bottom wall. These have been scratched out and cannot now be counted for the whole length of the wall. It is clear, however, that opposite the columns were two 'ticks', showing that the artist's original intention had been to furnish the hall with square columns. The ostrakon measures 9½ by 7½ inches.[1]

The second (Fig. 52) is of rather less interest, and represents the double door of a temple or shrine, fitted with its pivots and bolts, approached by a double stairway. On the left the artist or some other person has sketched a royal head. It measures 16 by 11 inches.

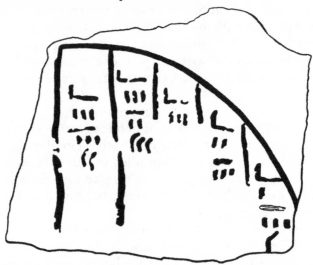

Fig. 53. An architect's diagram, defining a curve by co-ordinates. Probably IIIrd dynasty. Saqqâra. (From *Annales du Service*, xxv, p. 197.)

An architect's diagram of great importance has lately been found by the Department of Antiquities at Saqqâra (Fig. 53). It is a limestone flake, apparently complete, measuring about 5 × 7 × 2 inches, inscribed on one face in red ink, and probably belongs to the IIIrd dynasty. In the five spaces formed by the curve and the vertical lines are lineal measures expressed in cubits, palms, and digits (p. 63), as follows:

  1. '1 cubit, 3 palms, 1 digit' (41 digits).
  2. '2 cubits, 3 palms' (68 digits).
  3. '3 cubits' (84 digits).
  4. '3 cubits, 2 palms, 3 digits' (95 digits).
  5. '3 cubits, 3 palms, 2 digits' (98 digits).

[1] For a fuller discussion of this ostrakon see ENGELBACH, *Annales du Service*, vol. xxvii, pp. 72–5.

It seems clear that each measurement refers to the height of the vertical line to the left of it, and that these lines must be taken as all rising from the same datum level. The vertical lines will thus be offsets or co-ordinates which by their lengths determine the positions of a series of points on the curved line. Obviously, for this diagram to be of use, the distances of the vertical lines from one another (the ordinates) must be known. These data, however, are not given. It is natural to suppose that the vertical lines are intended to be equidistant, and the very fact that the distance is not specified makes it most probable that it is to be understood as one cubit, an implied unit elsewhere (*Journal of Egyptian Archaeology*, xii. 134). If a curve be

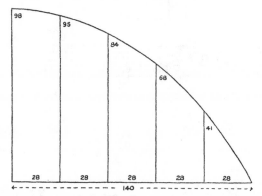

Fig. 54. Scale-drawing of the curve whose particulars are indicated on the ancient diagram shown in Fig. 53.

drawn to these ordinates and co-ordinates and made to drop to zero on the right, it will be seen that it is very similar to that of the ostrakon, except that it drops less sharply than the latter between 41 and 0.[1] The curve does not seem to be part of an ellipse.[2] This ostrakon was found near the remains of a solid saddle-backed construction, the top of which, as far as could be ascertained from its half-destroyed condition, closely approximated to the curve obtained from the data on the ostrakon (Fig. 54). It seems very likely indeed that the diagram on the ostrakon was intended to serve as a guide to the masons in constructing the saddle-back. If this be true, it shows that the system of drawing a curve by co-ordinates was understood.

[1] For a full description of this interesting ostrakon, see Gunn, *Annales du Service*, vol. xxvi, pp. 197–202, from which the above notes are partly transcribed.

[2] The question whether the Egyptians knew the method of tracing an ellipse by following round with a pencil, or its equivalent, held inside a string attached to two given points (the foci), keeping the string taut all the time, is discussed by Daressy in the *Annales du Service*, viii, pp. 237–41.

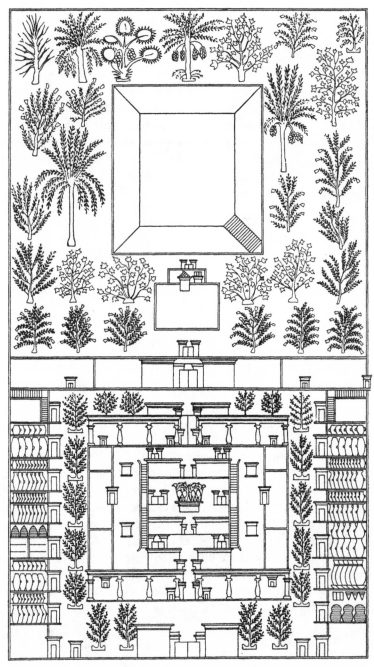

Fig. 55. Plan of an estate, from the XVIIIth dynasty tomb of Merirē‘ at El-‘Amârna. (After PERROT & CHIPIEZ, *L'Art Égyptien.*)

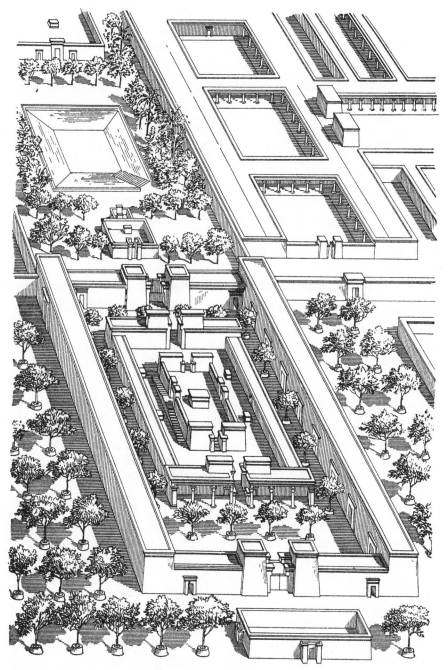

Fig. 56. Restoration of the estate of which part of the ancient plan is shown in
Fig. 55. (After Perrot & Chipiez.)

Since it is not truly to scale, we can infer that the curve was originally drawn by the architect to the form which pleased him, most probably on papyrus, and that he scaled off the co-ordinates from that and transferred his diagram roughly on to a flake for the instruction of the masons.

In many of the tombs of the nobles at El-'Amârna[1] it was the custom to have plans of the royal palace, and sometimes of private estates, sculptured on the walls; one of the best is seen in the tomb of Merirē' (Fig. 55). The appearance of the estate can be restored from it with tolerable accuracy (Fig. 56).

A plan of an estate, which possesses considerable interest, was recently bought from a dealer in Luxor. Under the title of *An Architect's Plan from Thebes*, it is described by Mr. N. de G. DAVIES,[2] who examines the question whether it was drawn to scale, and is of opinion that it was not. He remarks:

'The broad expanse of water at the bottom of the picture (Fig. 57) suggests either a canal or the River Nile in good flood, the space between this and the thin parallel line representing the sloping bank up which the rising waters will extend. Between this and the main boundary or river wall is another strip 32 ells (cubits) deep, planted with a double row of trees behind a low wall. . . . The entrance to the estate comprises a square enclosure, the centre of which is occupied by a tank 29 ells long and 23 ells odd broad. It has sloping sides down which and up which again half a dozen steps lead in a direct line from the entrance. The tank leaves ten ells' breadth round it within the enclosure, and this space is planted with a row of trees. The court is entered on the river side from a forecourt 21 ells 4 palms long, which extends along the bank almost to the water's edge. In the gateway to this a pedestal is set on which a burden or sacred bark might be deposited preparatory to carrying it down the steps. . . . The open lines which form the walls of the forecourt may indicate stone-work, in contrast to the solid black boundary wall of crude brick.'

Another drawing, worthy of consideration, is a map on papyrus made for the use of gold-miners of the time of King Seti I (Fig. 58). It represents two parallel valleys among the mountains with a winding valley connecting them. The entrances to four galleries are shown, together with a water cistern and a stela of the king close by. At the top right corner are the houses of the miners. The original is now in Turin.[3]

When the area in front of the southern colonnade of the XIth dynasty temple of El-Deir el-Bahari was being cleared by the Metropolitan Museum of Art, New York, circular tree-plots were found arranged in

[1] DAVIES, *El-Amarna*, i, Pl. XXXII; see also Pls. XXV and XXVIII; vol. iii, Pl. XXX, and vol. vi, Pl. XXVIII.

[2] *Journal of Egyptian Archaeology*, vol. iv, p. 194.

[3] LEPSIUS, *Auswahl*, Pl. XXII.

rows and containing, in most cases, the stumps of the ancient tamarisk trees. About this time a limestone ostrakon was found which was clearly the plan or project for laying out the grove (Fig. 59). Mr. H. E. WINLOCK, who directed the excavations, describes it as follows:[1]

'First, we must remember that drawings to scale were practically unknown to the

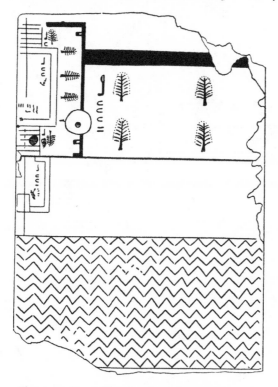

Fig. 57. An architect's plan of an estate, on a wooden panel, from Thebes. (From the *Journal of Egyptian Archaeology*, iv, p: 194.)

Egyptians, who were careless even of proportions. We need not be surprised to find, therefore, that the square temple and the ramp leading to it are represented by a mere symbol laid out on a centre line. . . . To right and left dots are laid out at the intersection of ruled lines. To the left we find three long rows of seven dots each—the tamarisk grove already excavated—but a closer examination of the stone shows a fourth row erased. Now it is an important fact that the left, or southern portico (Fig: 60), is shorter than the right, and it is easy to see what has happened.

[1] *Bulletin of the Metropolitan Museum of Art; The Egyptian Expedition*, 1921, 1922; p. 26.

The old landscape architect has paced off the length of the right-hand portico and found that he could work in four rows of trees. Then he has gone into the temple,

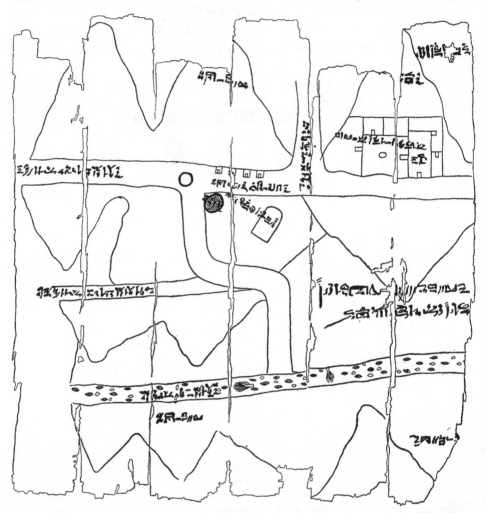

Fig. 58. Map on papyrus for the use of gold-miners of the time of King Seti I. It represents two parallel valleys among the mountains with a winding valley connecting them. The entrances to four galleries are shown, with a cistern and a stela of the king. (Now in Turin Museum.)

and squatting down on the floor has laid out a symmetrical design with four rows on both sides, which has stood until some more observant colleague has pointed out his mistake in supposing that both colonnades were of the same width, and he has scratched out his fourth row on the left. . . .'

It can hardly be doubted that the Egyptians, in addition to plans drawn on stone and on papyrus, made use of scale models. For certain details of their crafts, such, for example, as the determination of the point of balance of an obelisk, it would have been their only method of obtaining a solution, since their mathematics were very primitive.[1] In the Cairo Museum there

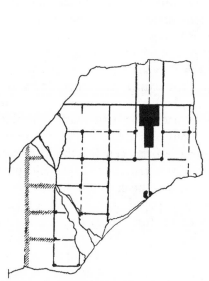

Fig. 59. Landscape architect's project for a grove of trees in front of the XIth dynasty temple at El-Deir el-Bahari. (From the *Bulletin of the Metropolitan Museum of Art*, New York, 1921–2, p. 27.)

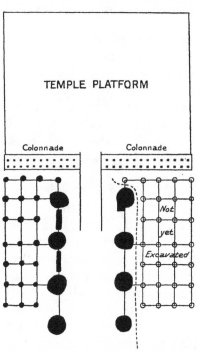

Fig. 60. The temple grove of the XIth dynasty temple at El-Deir el-Bahari as found. (From the *Bulletin of the Metropolitan Museum of Art*, New York, 1921–2, p. 27.)

are many models of gateways, towers, pylons, and other pieces of architecture, but none that can be definitely called an architect's scale-model. A model, reputed to have been found near Cairo many years ago, and rarely referred to in text-books, may, however, have fulfilled that purpose. It has been described as follows:[2]

'The model is of granite; it is 44·25 inches long, 34·65 inches wide, and 9·25

[1] See Chapter XX.   [2] GORRINGE, *Egyptian Obelisks*, p. 70 and Pl. XXXII.

inches deep. It shows a double flight of steps ascending to the level of the sanctuary. On either side of these steps are, first, sockets in which were formerly set models of the great sphinxes guarding the entrance. At the top of the steps are again, on either side, sockets for two smaller sphinxes. Beyond these are marked the positions of the two great pylons . . . and on the inner sides of these pylons are seen holes marking the place of the double gate of the sanctuary. . . . Farther on are shown the positions of the great walls enclosing the sanctuary.'

The model appears to date to the time of Seti I.

The commencement of a temple in ancient times appears to have been attended with many formalities, the most important of which seem, if we are to believe the sculptures, to have been carried out by the king himself. In the late temples, built when Egypt was under Ptolemaic or Roman rule, it can hardly be believed that the kings or emperors came to the temple in person and carried out all the details which they are shown in the sculptures as performing, though in dynastic times they may well have done so. It appears that the scenes, in such late times, had become merely traditional. Though scenes of foundation ceremonies are known in many temples of the dynastic period, the most complete examples are found in the temples of Edfu and Dendera. In these the king is first seen leaving the palace, preceded by the four standards of the primitive tribes of Upper Egypt; the Jackal of the First Cataract, the Hawk of Edfu, the emblem of Thebes, and the Ibis of Hermopolis. In turn we see him pegging out the limits of the temple area with Safkhet, the Goddess of History (Fig. 61), turning the first sod with a hoe before the god of the temple, throwing what may be seeds or grains of incense into the foundation trench, making a mud brick, and finally presenting the temple to the god. An Egyptian temple was not considered complete until it had been 'ornamented', that is, painted or sculptured throughout, but many years seem sometimes to have elapsed between the end of the building operations and the complete decoration of all the walls. At various stages of the painting and sculpturing, in the case of the temple of Edfu, the king formally presented the sculptures to the god, and in the scenes he is often depicted as offering the *kheker*—a hieroglyph meaning 'adornment'—to represent this stage in the proceedings. The inscriptions accompanying the foundation ceremonies are mostly very stereotyped, and are of value only in giving dates and the order of building.

Though the foundation scenes make no reference to them, objects were usually buried under and round temples and walls. These are now known as 'foundation deposits'. Sometimes they consist of animal sacrifices only; at other times a variety of objects was deposited, consisting of plaques inscribed with the name of the king, pottery, fayence and stone vases, and specimens of the tools with which the work had been carried out. Queen Hatshepsowet, in her temple at El-Deir el-Bahari, put in miniature copper tools, some with their wooden handles, brick-moulds (Fig. 263), 'rockers' (Fig. 89), mallets, hoes, &c., while in other temples the whole

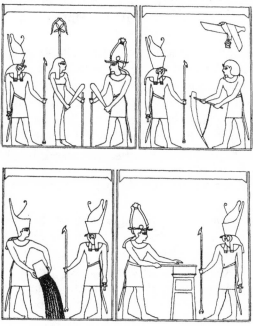

Fig. 61. Part of the foundation ceremonies performed at the temple of Edfu. The king is seen pegging out the limits of the temple with the goddess Safkhet, cutting the first sod, pouring seed or grains of incense into the foundation-trench, and moulding the first mud-brick. (From ROCHEMONTEIX-CHASSINAT, *Le Temple d'Edfou*, Pl. XI.)

deposit was made of rough red pottery. Good collections of foundation objects can be seen in the Cairo Museum. No fixed rule seems to have been observed for the position of the objects with regard to the temples and walls, though different periods show preferences. The most varied collections of objects are found in the New Kingdom. In the Temple of El-Deir el-Bahari not only were they put beneath the walls, but also in pits round the courtyard.

Although the Egyptians did not always take the trouble to make the foundations of their buildings level, they were capable, if they wished to do so, of levelling a large surface with great accuracy. No information has come down on the method employed, but it can fairly safely be assumed that they achieved their purpose by the natural method which the modern cultivator uses, namely, by taking advantage of the fact that the surface of still water is always at one level. In a building on alluvium, levelling presents no difficulties, since water channels can be run over the area to be levelled in as many places as is desired. To level a stone surface, however, would require certain modifications in this method. The most accurately levelled area of which exact data are available is the pavement on which the casing-blocks of the Great Pyramid rest. It does not extend under the whole pyramid, since it is known that a core of rock rises up into the mass of the building. How far the pavement is laid under the pyramid is not known, but outside the casing it extends for about two feet. A recent examination of the pavement by the Survey of Egypt showed that the visible parts lie on an almost perfect plane, though the whole plane slopes up from the north-west to the south-east corner to the extent of just over half an inch.

The Egyptian method of flattening a stone surface was to obtain, at different points upon it, small surfaces lying on the plane to which the whole area was afterwards to be reduced. Though the method of obtaining these 'facing-surfaces'[1] must have been different for various classes of work, the procedure of reducing the whole area to the plane, after they had been found, seems to have been by means of 'boning-rods', which are described on page 105.

The method of obtaining the facing-surfaces when levelling a pavement would probably have been to flood the area with water by banking up the edges with mud, and to cut down the rough surface at convenient places to a fixed distance below the surface of the water. In such a method certain precautions would have to be taken, such as making a final check of the facing-surfaces as nearly simultaneously as possible on a still day. In this manner the levelling of a surface of any size could be carried out to the limit of visual accuracy; it is only a question of patience and good organization of skilled labour.

The slight slope of the pavement of the Great Pyramid can conceivably be accounted for if we suppose that a north-west breeze was blowing at the time of the checking of the facing surfaces. This would be amply sufficient

---

[1] We have used the term 'facing-surfaces' to describe the small surfaces made on stonework to the plane of which the whole face of the stone was afterwards to be dressed.

to pile up the water towards the south-west and to produce the error observed.

The unit of measurement of the Egyptian in building was the Royal Cubit, which varies considerably in different dynasties and even in the same dynasty. It will suffice to give a few examples of the length of this unit:

Great Pyramid (after PETRIE) . . 20·62 in.
Royal Tombs at Thebes (after CARTER) 20·595 in.[1]
Quarries at Aswân . . . . 20·67 in.
Ptolemaic (cubit rods from Dendera) . 20·1 and 20·77 in.

The cubit was divided into seven *palms*, and these, in turn, were subdivided into four *digits*. All measurements of buildings given by the Egyptians in their writings are in these units.

Our knowledge of the value of the cubit is derived from actual cubit rods, both in wood and stone, which have been preserved; from dimensioned lengths and levels inscribed by the ancient masons on stonework and on quarry-faces, and from the measurement of the lengths, breadths, widths of doorways, &c., of finished buildings where, more often than not, even cubits were used.[2]

The common form of cubit rod was of rectangular section—often square—with one edge bevelled. It was divided into seven parts, the last, or the last two, being again divided into four. Often a line is found dividing the rod into halves, and occasionally there is an extra mark close to the fourth palm, which may be some kind of foot, but which cannot as yet be more definitely placed. Stone cubit rods of late date are known, some with the digits subdivided into halves, thirds, quarters up to sixteenths, some having what appear to be astronomical data on them, but these must be considered as ceremonial objects rather than the tools used by the masons. Two stone cubit rods of Roman date from Dendera have their faces subdivided into fourths, fifths, sixths, and sevenths respectively.

It is not known whether the ancient Egyptians had the grotesque mixture of units used by the modern Egyptians in their trading, such as the span, the clenched fist with the thumb extended, and so forth.[3] It is likely, however, that many foreign and local measures were in use at the same time as the Royal Cubit, but it is extremely doubtful if these were employed in architecture. Writers have tried to prove that the uneven heights of the

[1] *Journal of Egyptian Archaeology*, iv, p. 149 (converted from metric).
[2] The method of obtaining ancient units by this means is described in PETRIE, *Inductive Metrology*.

[3] Cubit rods of 26·8 inches, divided into 7 palms, of the XIIth dynasty are known (PETRIE, *Weights and Measures*, p. 40).

courses in the Great Pyramid are due to the presence of foreigners who were employed on its construction and who used their own units.[1] The fact that the *top* of a course was flattened or levelled after it was laid (p. 100) by itself rules out this theory (apart from its improbability), since the course would have to be reduced to the height of the lowest block in it, and the result would not be likely to represent any particular unit. Though the work in the granite quarries at Aswân was divided into what appear to be foot-lengths, this should not be taken as proof that the foot was an Egyptian measure, since it is possible that what has been called the 'foot' may have been what had been found by experience to be the minimum length in which a workman could carry out his poundings (p. 26). At

Fig. 62. Field-surveying, from the tomb of Amenhotpe-si-se at Thebes. (From DAVIES, *The Tomb of Two Officials of Tuthmosis IV*, Pl. X.)

any rate, the Egyptian makes no reference to a foot. As for the 'Pyramid Inch' it has no existence outside the minds and writings of those technically known as 'Pyramid Cranks', who require it to help out their fantastic theories on the prophecies supposed to be embodied in the lengths of the passages and other dimensions in the Great Pyramid.

The determination of the Egyptian units of length from the measurements of different parts of buildings is, at best, a rather unsatisfactory undertaking, except when the determination is only required very approximately. The faces of Egyptian masonry were left rough and dressed afterwards, which would not conduce to leaving a finished face measurable to an exactly even unit. In medieval masonry, on the other hand, such a method may be extremely valuable.

The only records of the means used by the Egyptians for surveying are found in the tomb-scenes and in certain statuettes. From these it appears that a cord was used in land measurement. In a scene from the tomb of

[1] See TARRELL, *Ancient Egypt* (1925), p. 36.

Fig. 63. Basalt statuette of the scribe Penanhūret, holding the royal surveying cord, tied with the cachet of the god Amūn. (Photograph by the Survey of Egypt)

Amenhotpe-si-se (No. 75), at Thebes,[1] men can be seen measuring the standing corn belonging to the estate of the god Amūn (Fig. 62). Although they are not shown in this scene, it appears from similar scenes in other tombs[2] that the cord was furnished with knots at regular intervals. In order, perhaps, to show it was a standard measure, the cord had a tying-up string furnished with a uraeus-crowned head of a ram, this probably being the cachet of the god. When the measuring-cord was in use the cachet was attached to the shoulder of one of the men holding it. In a statuette of the scribe Penanhūret, in the Cairo Museum, he is seen holding on his knees the royal surveying-cord which is also furnished with the cachet (Fig. 63).

The unit of area for measuring land was the *stat*, the Greek 'aroura', which was equal to 1 square *khet* or 100 cubits squared (2735 sq. m. or about $\frac{2}{3}$ acre). The *stat* was divided into 2 *remen*, 4 *heseb*, or 8 *sa*. Other subdivisions of the *stat* were expressed in terms of the *cubit of land*, i.e. a strip of land 100 cubits in length by one cubit in depth ($\frac{1}{100}$ *stat*). A *thousand of land*, or *kho*, was equal to 10 *stat*.

The unit of capacity for corn was the *hekat* of 292 cubic inches (4·785 litres or 8·42 pints), which was divided into 10 *henu* ('hin') and also sub-divided as far as $\frac{1}{64}$. The double- and quadruple-*hekat* are also known as measures. Another unit of capacity was the *khar*, which was equal to 5 quadruple *hekat* or $\frac{2}{3}$ of a cubic cubit. It is possible, however, that masonry was measured by the cubic cubit only.

Though no measuring-cord has been preserved from ancient times, a mason's cord-and-reel of the XIth dynasty is known (Fig. 266). The cord is wound on a split reed, which revolves round a spindle, the latter being in one piece with the handle.

Having briefly reviewed the systems of measurement of the Egyptians, together with the little that is known of their methods of measuring, we have to consider how accurately they could survey the limits of a temple or other construction. The Great Pyramid appears to be by far the most accurately set-out monument in Egypt, and, until quite recently, it was thought that no side of the base differed from another by more than 1·8 inches. This, on a side of some 9,000 inches, means an accuracy of 1 in 5,000. Since steel-taping, with suitable corrections, can hardly attain this degree of accuracy, students of the Great Pyramid were somewhat nonplussed to account for such precision. Recently, however, the lengths of the sides of the Great Pyramid have been determined by the Survey

---

[1] DAVIES, *The Tombs of Two Officials of Tuthmosis IV*, p. 11 and Pl. X.

[2] COLIN CAMPBELL, *Two Theban Princes*, p. 86, and BORCHARDT, *Aegyptische Zeitschrift*, xlii, p. 70.

of Egypt by finding a number of points where the edge of the casing-blocks met the pavement; and by plotting these points the true azimuth of each side has been determined.[1] The results show that the sides of the Pyramid base had the following lengths:

North	.	.	.	.	.	9065·1 inches
East	.	.	.	.	.	9070·5 ,,
South	.	.	.	.	.	9073·0 ,,
West	.	.	.	.	.	9069·2 ,,
				Mean		9069·45 ,,

It will be seen from this that the difference between the longest and the shortest sides is 7·9 inches, which is more easily explicable than the result from the earlier survey. Whether such accuracy could be obtained by means of a measuring cord cannot be determined without personal experiment. In the opinion of the writer it could, given a flat, level surface. Even if the cord, on experiment, could not be made to give this accuracy, there is no need to assume that any complicated apparatus was used. The old method of putting measuring rods alternately end to end could give an accuracy far superior to that of the Great Pyramid. The original triangulation base for the Survey of England on Salisbury Plain was set out by means of rods, though there were, of course, many corrections applied and precautions taken which the Egyptians cannot be expected to have known. With modern corrections, rod-surveying can be done to an accuracy of 1 in 250,000. Using two model cubit rods of hard wood, one of the writers was able to set them end to end twenty-five times and return along the same line successively to the starting-point without any discrepancy which could be measured. It must be remembered, however, that there is *no proof at all* that the Egyptians measured by putting rods end to end in the manner described.

As regards the means used by the Egyptians to obtain a good right angle one has to rely on supposition, but it is unjustifiable even to hint that any instrument in the nature of a theodolite was used as long as simple appliances can be made to give the maximum accuracy observed. From the recent determination already referred to, it seems that the angles of the Great Pyramid base had the following values:

North-east	.	.	.	$90° + 3'2''$
South-east	.	.	.	$90° - 3'33''$
South-west	.	.	.	$90° + 0'33''$
North-west	.	.	.	$90° - 0'2''$

---

[1] Cole, *Survey of Egypt Paper* 39; Borchardt, *Längen und Richtungen der vier Grundkanten der* *grossen Pyramide bei Gise*; Engelbach, Précis of Survey Paper 39, *Annales du Service*, xxv, p. 167.

There is thus a maximum error from the right angle of 3¼ minutes of arc. In the case of the Great Pyramid, the square could not be corrected by measuring the diagonals, since a core of rock rises into the mass of the building. This practically rules out the possibility that the right angle was set out by a triangle having its sides proportional to, say, 3, 4, and 5, or 5, 12, and 13, which the Egyptians may well have known.

A right angle can be set out with fair accuracy by purely visual methods,

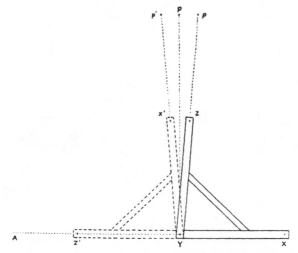

Fig. 64. Diagram to illustrate the principle on which a
right angle may be set out by visual methods.

though exactly what form the sighting-appliance had in ancient Egypt we have no means of knowing. The instrument made by one of the writers to determine what accuracy could be obtained by visual methods consisted of two pieces of wood about 7 feet long set roughly at right angles to each other, with pins put at the end of each arm and at the joint. The method by which a right angle was obtained to a line *AY* was as follows: the appliance, which was similar to that shown in the illustration (Fig. 64), was placed so that the pin at the angle was exactly over the point *Y*, and the line *XY* sighted in line with *AY*. Sighting along *YZ*, a peg was put in about 100 yards away at *p*. The appliance was then turned through a right-angle, the pin at the angle being kept over *Y*, and the pins *YZ* were sighted so as to lie along *AY*. Another peg *p'* was placed beside *p* by sighting along *YX*, and the distance between *p* and *p'* was bisected (at *P*) and the right-angle (*AYP*) thus obtained was checked by means of a theodolite. It was found that in no experiment did the error from the right angle obtained by this visual method exceed 1½ minutes. It is possible that the appliance actually used by the pyramid-builders was of a nature akin to the Roman *groma*,[1] which consisted of two pieces of wood attached in the middle to form a cross, from the four arms of which were suspended plumb-lines, the whole being in turn suspended from the middle of the cross. The principle of the *groma* is the same as that in the visual method described. Whether such an appliance, being thus suspended, would be steady enough to permit of very accurate readings being taken may well be doubted.

Though, in general, the temples were not oriented with any very great accuracy, the pyramids, and especially the Great Pyramid, were oriented very closely to true North. Since we have no reason to believe that the Egyptians used Polaris, it is probable that they sighted on a star which only sets for a few hours, and bisected the angle between the setting position, the observer, and the rising position. Again, it is possible that they took their observations on the extreme positions of a circumpolar star. It seems that by such observations it is possible to obtain an orientation as close as that of the Great Pyramid, neither the east nor the west side of which is closer than 2 minutes 30 seconds to true north. The actual orientation of the four sides is as follows:[2]

West side	.	.	.	.	2′30″ W. of N.
East side	.	.	.	.	5′30″ W. of N.
North side	.	.	.	.	2′28″ S. of W.
South side	.	.	.	.	1′57″ S. of W.

---

[1] For information on this appliance, see *Discovery*, Sept. 1925, and SLOLEY, *Ancient Egypt* (1926), p. 65.

[2] The subject of the orientation of the Pyramid is discussed at length in BORCHARDT, *Längen und Richtungen der vier Grundkanten der grossen Pyramide bei Gise*, p. 9, where other references to orientation may be found. It may also be observed that the present orientation of the Great Pyramid may well be considerably different from that when it was constructed, since the direction of True North is likely to have changed during nearly 5,000 years.

VI

FOUNDATIONS

EGYPT presents certain peculiarities of soil which are not common to other countries; the absence of rain and the consequent general dryness give the builder great advantages, and the dry, hard rock or rocky debris, found almost everywhere, makes a firm and suitable basis on which to build.

On the other hand, the alluvial soil which forms the level floor of the Nile valley offers some peculiar difficulties to the builder. When the alluvium is dry, it is capable of bearing immense weights, but with the rise of the Nile it becomes soft and yielding—a very treacherous surface indeed. At low Nile, when its surface is baked by the summer sun and it becomes thoroughly dry, the mass contracts, and wide and deep cracks are formed in all directions.

The annual deposit of sediment by the rich water of high Nile has slowly elevated not only the surface of the valley-floor but also the level of the river-bed. The bottom of the trough in which the Nile rises and falls is itself raised. In consequence, many buildings which, when they were founded, stood well above the level of high Nile, are now inundated every season. The floor of the temple of Ptah at Memphis is more than eight feet below what is now arable land, and is flooded every year. In the Delta the ancient temples often lie at a far greater depth than this. The forecourts of the temple of Abydos and of the Ramesseum at Thebes have suffered badly from the sinking of the ground under the immense weight of the masonry, and the temples of Karnak, in places, used to stand several feet deep in water each year, though in the last case elaborate precautions are now being tried in order to keep them dry all the year round. The huge funerary temple of Amenophis III, which stood behind the colossi of 'Memnon' at Thebes, has now almost completely disappeared, though there is contemporary evidence to prove that it was once one of the largest and most splendid of all the temples in Egypt.[1] Yet the site was chosen with so little foresight that, well under two hundred years from its completion, the flooding of its foundations had rendered it in such a state that it was apparently only fit to serve as a quarry for the new temple of King Meneptah. At present what foundations remain are six feet or more below cultivation

[1] BREASTED, *Ancient Records*, ii, § 904.

level. The colossi alone have escaped the general destruction. It may be remarked that one cannot always put down the destruction of a temple to the action of water or to bad foundations; a personal dislike of the founder by a subsequent king must always be taken into account. The temple of Meneptah, which was built from the ruins of the temple of Amenophis III, was itself almost completely destroyed, though it has always stood well above flood-level.

The rise in level of the Nile is a thing so obvious that it is impossible to believe that it had not been to a certain extent studied by a people so advanced as the Egyptians; yet in many places new monuments and temples were set up, or old ones rebuilt, at a level so little above the high Nile of the period that a few moments' thought would have foreseen their inevitable fate.

An interesting example of the accumulation of the alluvial soil can be studied at Hieraconpolis—the ancient *Nekhen*—which shared with *Nekheb* (now El-Kâb) the honour of being the frontier city of the south in the earliest dynastic times, if, indeed, it did not exist before that. When the site was excavated,[1] vertical measurements were taken from a datum line established just above the level of the highest part of the temple ruins. At a depth of 13 feet 2 inches below datum, the lowest level was found, which was a rough stone facing to a sandy mound. This mound was evidently the site of a holy place. The desert surface lay a good three feet below these stone faces. What type of building originally stood on this mound it is impossible to conjecture, but at a depth of 6 feet 7 inches below datum there are foundations of a temple which is more or less dated by the well-known copper statue of Pepi II now in the Cairo Museum. It is certain that, when this mound began to grow, the desert level was well above high Nile. As this site must have been under observation from the earliest years, the gradual encroachment of the Nile must have been sufficiently evident, yet it seems that the place continued in use, and that a temple, with crude brick foundations, was built at a period when the certainty of a further rise of the Nile was beyond question. It must be borne in mind, however, that it was not entirely due to lack of observation that a site was adhered to in spite of the rise of the Nile. Most of the primitive gods seem to have been water-deities, to judge from the fact that their images were represented on the sculptures as mounted on boats when they stood on the altar. A holy site, once established, could not well be changed.

It is extremely difficult to gauge the proximity of an ancient site to the

---

[1] QUIBELL and GREEN, *Hieraconpolis*, Part II.

Nile at any given period. For example, in the *Description de L'Égypte*, Plate 5, it can be seen that the main channel of the Nile passes on the west side of what is known as the 'Gezîra' at Luxor, which is now dry most of the year, and, from the rather vague description of Pococke in 1772, it passed still farther west in his day. There is evidence to show that the Nile at Luxor in dynastic times has shifted its course more than once, and that it has at times passed close to the foot-hills opposite the valley leading to the Tombs of the Kings.

The earliest buildings which have survived are constructed on the hard desert surface, since those built elsewhere have been swallowed up by the rising Nile bed. It is therefore not known if there was any difference between the foundations of the ancient buildings constructed on alluvium and those of the later buildings. The most convenient monuments of the Old Kingdom whose foundations can be studied are the pyramids and mastabas of Gîza. The procedure in laying the first course of blocks was to recess the rock so that, when the blocks were laid, the rough top surfaces of a series of them—often the whole course—should be all more or less at one level. The top of the course was dressed after the blocks were laid. In the case of a pavement the same system was used, the surface being dressed last of all. It will be seen (Chapter IX) that this is the fundamental principle of Egyptian building practice, the blocks being always laid with the least possible amount of dressing on them.

Advantage was often taken of this method of laying blocks when the desert on which a wall was to be constructed was sloping, the foundation trench having the appearance of a series of steps. Sometimes, when the surface of the desert was sloping back from the line of the proposed wall, the first line of blocks was laid on beds which also sloped back at an angle. This can best be seen in the Queens' pyramids to the east of the Great Pyramid. A series of blocks, with beautifully fine rising joints between them, can often be seen lying on beds which slope back at different angles to the face of the wall. This method was an alternative to filling, and undoubtedly more solid in such examples as the casing-stones of a pyramid or mastaba.

When the surface of the ground was of such a nature as that on which the temples of El-Deir el-Bahari at Thebes were built, there was little need for elaborate foundations. These temples stand on the detritus from the limestone cliffs above them. The bottom of a trench, a yard or so deep, was neither better nor worse than the surface of the ground. Further, there was nothing to fear from the rise of the Nile, and rain is a great rarity; therefore

they were built with practically no foundations at all. Their ruin is not due
to the foundations, but to bad workmanship, which is worse in the XVIIIth
than in the XIth dynasty temple.

Another example of building on such a surface is that of the little temple
of Tuthmosis III at El-Kâb. Here the foundations are of the most in-
different kind. The ground consists of soft disintegrated sandstone debris,
and it was trenched to a sufficient depth to receive one layer of small blocks
of stone. This was all the preparation that was made; nevertheless, as long
as water did not make its way into the subsoil, it was sufficient. The building stood nearly perfect until early in the last century, when a Pasha pulled
it down to use the stone for a sugar mill! Standing, as it did, barely five
yards high, it was only a simple construction, but the same indifference to
the adequacy of foundations is also found in larger columnar buildings, and
on ground which was manifestly not too firm.

The best example of a temple built on alluvium is that of Karnak. The
site must have been a holy one from early times, but those who were
responsible for building the small shrines in the XIIth dynasty could hardly
have foreseen the colossal series of structures, occupying some 600 acres, to
which the little shrine would eventually give rise. All the buildings of
Karnak have one thing in common: their foundations are poor, and, since
they rest on alluvium, the gradual infiltration of water into the subsoil, and,
until recently, on the surface, has made many of them extremely insecure.

The method of making a foundation in the alluvium at Karnak was
generally a very simple one. A trench of the necessary width and length
was dug and the bottom of the trench was filled to a depth of some eighteen
inches with dry sand.[1] A very level surface could thus be obtained on

[1] In rebuilding on the site of an older construction,
however, the Egyptians often did not trouble to expose the old sand-bed, thinking, perhaps, that the
chips and rubbish made as good a foundation. The
late M. Legrain, Director of Works at Karnak,
cleared the south-west corner of Pylon VIII to
observe on what layers of materials it rested. He
gives the following figures (*Annales du Service des
Antiquités*, iv, p. 23):

Composition of Soil	Thickness of Layer (*metres*)	Level at Bottom of Layer (*metres*)
Ground level	—	0·00
Stone foundations	0·70	0·70
Masonry chips and rubbish	0·20	0·90
Soil containing flints	0·50	1·40
Earth	0·80	2·20
Flints, ash, pottery, and bone fragments . .	0·20	2·40
Earth	0·30	2·70
Sand	—	—

Fig. 65. Foundations of a column (no. 24) in the north-east part of the Hypostyle Hall at Karnak, (From the *Annales du Service*, Vol. XXIII)

Fig. 66. Present aspect of the pylon of Ramesses I at Karnak

which to lay the first course of the masonry. Sand thus used, when care is taken that it cannot escape laterally, is one of the best materials on which to build. It was not here that the ancient architects built badly; it was rather their habit of putting, between the sand and the huge blocks forming the walls and columns, a few courses of small, friable stones (Fig. 65). Why this was done it is extremely difficult to say; it might be imagined that it was in order to economize in the large blocks, since they would not show beneath the ground, but this hardly explains it, since practically the whole

Fig. 67. Foundations of the pylon of Ramesses I (No. II) at Karnak.
(From a photograph by the late M. G. Legrain.)

surface of the walls and columns was covered with gesso and painted, thus covering all the joints.

Any blocks seem to have served for the foundation of a pylon; for example, in the consolidation work by the Antiquities Department on Pylon III (of Amenophis III) at Karnak, fine alabaster blocks were found bearing scenes and inscriptions of Amenophis I.[1] These blocks have since been re-assembled and form no inconsiderable part of a chapel. The foundations of Pylon II (Fig. 67) consist of wretched little blocks, totally inadequate to support the immense weight of the pylon, which had been taken from a destroyed temple erected by the heretic king Akhenaten. These blocks do not even extend beyond the base of the massive wall which rests on them. The effect has been that they have become crushed under the weight, and the large blocks resting on them have, in consequence,

[1] PILLET, *Annales du Service*, xxii, pp. 238–40; xxiii, p. 112.

split. This pylon has indeed suffered (Fig. 66); its collapse has been due not only to its internal weakness, but to one of the governors of Luxor using it as a quarry, gunpowder being employed to help to extract the blocks.

In the Festival Hall of Tuthmosis III at Karnak the foundations are of the same poor quality (Fig. 68). The set of small blocks of which they were composed was laid at right angles to the length of the wall (i.e. as 'headers'). The settling of some of these small blocks has resulted in the cracking of the large blocks in the upper courses.

If the foundations of the pylons and walls were bad, those of the columns

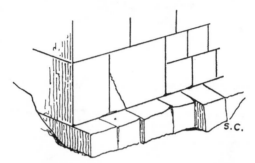

Fig. 68. Foundations of the Festival Hall of Tuthmosis III at Karnak, showing how they were laid as 'headers'. Being of insignificant size, the setting of one of them often resulted in the breaking of one of the large blocks above.

were worse. A hole was dug in the alluvium, not so that the foundations should be wider below than at ground level, but sometimes the reverse! Small stones, roughly rectangular, measuring some 20 × 10 × 8 inches, were laid in a pit containing a bare 18 inches of sand, the pit being often wider at the top than below. On this insecure base rested the great pair of semicircular half-drums on which the giant column was erected. It can easily be imagined that the constant flooding of the subsoil did not improve the stability of the columns in the Great Hypostyle Hall. Before the late M. George Legrain became Director of Works at Karnak, one of the columns in the northern portion of the hall had declined very perceptibly from the vertical, and it was decided to take it down drum by drum and re-erect it on a more secure foundation. It was then that one of the writers was enabled to study not only the foundations but the manner in which the various parts were mortared together. The re-erection was duly carried out and the column made secure. Lately, the foundations of many more columns have been consolidated without taking them down, the column being strutted and the foundations replaced by concrete, half being done one year and the remainder the next.[1]

On the 3rd October 1899, a most unexpected calamity befell the Hypostyle Hall.[2] For centuries the rising waters of the Nile had not flowed into the ruins of Karnak, but had entered by infiltration. The Nile water itself would do comparatively little harm to the masonry, but the water rising through the soil, charged with salts, was having a most disastrous effect, converting the stone, in places, into a fine floury dust. It was decided that the best course would be to open a free passage for the Nile and thus let in clean water. On the above-named day, eleven columns in the north part of the hall fell down, the noise being heard in Luxor, some two miles away. In addition to this, many other columns were shaken and partially dislocated. All the columns, for some unknown reason, fell towards the west (i.e. riverwards), and it was a strange sight to see the huge masses tilted over from the very base. The insecure foundations had given way, and it is remarkable how insufficient some of them had been for their purpose. The columns, falling, pivoted on their bases, and the large, flat base-drums on which they stood were driven into the ground, the upper drums falling apart and lying in long rows. Fortunately, none of the greater columns in the central rows of the hall fell over, but they did not escape the accident by reason of their better foundations, for they also stand mostly on comparatively poor little blocks, carelessly placed in a hole. An examination of the interior of the holes which contained the foundation of the fallen columns showed that the action of the salty infiltration water had reduced them to such a state that the slightest friction would turn them into powder.[3]

Recent excavations by the Antiquities Department round the bases of the columns of the Hypostyle Hall have shown that at some period in their history an attempt had been made to fill up the space between the foundations of some of the columns. It seems, however, that this work was never completed.

The superb column of King Taharqa (Fig. 69), the last survivor of ten, was leaning in such a dangerous manner towards the north-east, that it has

[1] Pillet, *Annales du Service*, xxiii *et seq.*, where the procedure is described.
[2] For an account of the accident and its consequences, see *Annales du Service*, i, p. 121.
[3] As an experiment, a large trench, 7 metres deep,

has been dug round the main group of temples, and drainage-pipes have been laid within the temple area in the hope of preventing further infiltration. It remains to be seen whether this will have the desired effect.

recently had to be taken down block by block and rebuilt on secure foundations.[1]

The foundations of the obelisks were very little better than those of the pylons and columns. Though a couple of them were removed to Rome in late times, all the remainder—at least eight—have fallen except two; those of Queen Hatshepsowet and of King Tuthmosis I. The last has a very decided lean riverwards (Fig. 70) and it seems to be only a question of time until it also crashes down. Nothing can be done to bring it back to the vertical, as it is badly cracked in the middle. The bases of the fallen obelisks all show a lean in the same direction, and though earthquakes and the Assyrians may have played their part in the overthrow of the Karnak obelisks, most of the damage must be attributed to bad foundations, which consist of a few courses of fairly well squared sandstone blocks resting on a metre of sand (Fig. 71). As long as the alluvium remained dry, this was sufficient, wonderful as it may seem, seeing that they were upwards of 60 feet high and weighed more than 250 tons. With the flooding of the subsoil, however, the powdering of the foundation blocks rendered the foundation hopelessly inadequate.

At or about the XXVth dynasty, a change took place in the laying of foundations, this period also showing a marked reaction, an effort to return to better forms and better sculpture. It was then discovered that it was well to have some regard to the foundations of a building. Advancement would not be expected from Ethiopia, yet under the princess of that country the buildings show a prodigality of materials in the foundations. Far from making a few trenches and holes to receive the walls and columns, the whole area of the temple was completely covered with carefully laid blocks three or four courses deep. This method continued in use until Roman times, and was made use of however small the building was. The temples of Dendera, Edfu, and Kôm Ombo are all constructed in this way. The large temple of Nectanebos II at El-Kâb, which stands on a slight hillock, has, where the ground slopes downwards, foundations of not less than eight courses of stone, forming a platform on which the structure rests, and the little temple of the same king outside the main gate at El-Kâb has a platform of large squared stones, though it consists of but one apartment surrounded by a colonnade. At Kôm Abu Billo, in the Delta, a small temple has been built in late times in the accumulated town rubbish, far above the highest water-mark; yet it has for foundations a platform more than nine courses deep (Fig. 72).

[1] CHEVRIER, *Annales du Service*, xxviii, p. 120.

Fig. 69. Column of King Taharqa at Karnak, leaning from the vertical owing to the settling of its foundations. It has recently been taken down block by block and rebuilt

Fig. 70. Obelisk of King Tuthmosis I at Karnak, showing its present deviation from the vertical owing to the settling of its foundations

Fig. 71. Sandstone foundations of the pedestal of the fallen obelisk of
Tuthmosis III, before Pylon VI; Karnak

Fig. 72. Remains of a small Roman temple at Kôm Abu Billo, showing the massive platform
on which it is built

It must not be imagined that the Egyptians were not aware of the frequent failure of their foundations. At a quarry near Gebelein there is a stela of the XXIst dynasty, where it is related that the king, when in his palace at Memphis, had a dream in which the god Thōut appeared to him and warned him that the Nile was attacking the foundations of the canal-wall of Tuthmosis III at Karnak.[1] The king thereupon ordered his engineers to take 3,000 men to obtain stone from Gebelein in order to effect repairs. Since nearly all the stone-work of Tuthmosis III at Karnak is sandstone from Gebel Silsila, it is not quite clear why limestone should have been chosen for the repairs, and where, if at all, they were carried out.

Before leaving the subject of foundations, it may be of interest to consider the quality of medieval buildings in Europe. There is a favourite saying in England that, in the old days, our forefathers knew how to build well, and did it. An examination of nearly all our great medieval buildings reveals to us that the foundations were mostly left out; indeed our ancestors did *not* build well as a rule. It is the thickness of the walls, the mere mass, that has enabled the buildings to stand as well as they have; their arcated style of architecture, moreover, is very flexible, provided that the arches are pointed and not semicircular. It was nothing unusual, however, for towers to fall down in a comparatively short time after they were built. Sir T. G. Jackson has shown how the buttresses in the north wall of Winchester Cathedral were merely stuck on without foundations. They were supposed to strengthen the walls against which they were placed; in reality, they were pulling them down. Since we, in medieval times, with not only the splendid example of the Romans before us, but also the experience of our civilized neighbours, so neglected our foundations on a soil that was continually flooded, we should be lenient in our criticism of the Egyptians. They had no such advantages.

[1] BREASTED, *Ancient Records*, iv, § 629.

VII

MORTAR

TO the modern mind, mortar plays a very important part in the construction of a building, whether it be of stone or brick, and in these days cement is of even greater importance still. This results very naturally from the fact to which attention has already been called, namely, that, with us, buildings are composed of stones of insignificant size, or of brick. The material which binds these small parts into a solid mass must therefore play a part of the greatest importance.

Where megalithic building was the custom, with the blocks weighing anything from five to fifty tons each, the cohesive power of the mortar was of very little importance. The component parts of the structure were held together by friction occasioned by dead weight.

The Egyptian masons did not regard mortar in any way as a cementing medium, and some of the most serious students of Egyptian methods of construction do not seem to have appreciated this fact. Sir (then Mr.) W. M. F. Petrie, writing on the casing-blocks of the Great Pyramid, made the following statement: [1]

'Several measures were taken of the thickness of the joints in the casing-stones. The mean thickness of the joints of the north-eastern casing-stones is 0·02 inches; and therefore the mean variation of the cutting of the stone from a straight line and from a true square is but 0·01 on a length of 75 inches up the face, an amount of accuracy equal to the most modern opticians' straight-edges of such a length. These joints, with an area of some 35 square feet each, were not only worked as finely as this but cemented throughout. Though the stones were brought as close as $\frac{1}{50}$ inch, or, in fact, into contact, and the mean opening of the joint was but $\frac{1}{100}$ inch, yet the builders managed to fill the joint with cement; despite the great area of it, and the weight of the stone to be moved—some 16 tons. To merely place such stones in exact contact at the sides would be careful work; but to do so with cement in the joint seems almost impossible.'

It was, however, the presence of the mortar in the bedding-joints which enabled the blocks to be laid. Without it, the Egyptians could not have laid them at all. Apart from its value in facilitating setting, mortar plays an equally important part in masonry. Since the lower courses of a lofty stone wall support a great weight, it is essential that the top of a course

[1] Petrie, The Pyramids and Temples of Gizeh (new ed. 1885), p. 13.

should evenly support the course above it, otherwise cracked blocks will result. A coat of hard-setting mortar will provide the much-needed level surface, distribute the weight, and preserve the integrity of the blocks.

The north of England, Devonshire, and the Scottish Lowlands are among the places where masonry and stone-setting with comparatively large blocks can be well studied. Mortar is made use of to form an even bed and to facilitate 'setting', which is the technical term for getting a block exactly into place; the mortar being, practically speaking, a lubricant. It is obvious that, unless a stone of considerable weight is laid on a bed of such a nature, so that it can be adjusted hither and thither, good setting cannot be obtained. For masonry of any fineness, a layer of some sort must be used having the consistency of butter—in other words, a lime-cream, which as a cement or mortar is without value. In the south of England, this creamy mixture is known as 'fine-stuff' or 'butter' and in the north as 'softening'.

Egyptian mortar consists of gypsum and sand, with a considerable admixture of impurities. Mr. A. Lucas, formerly Director of the Chemical Department, Egyptian Government, states that he has, as yet, found no lime in Egypt before Roman times. He suggests that the reason for this is that the lime-making requires a much greater heat than is required for converting gypsum into plaster, which was universally used for mortar in masonry, an economy in fuel being a consideration in a sparsely wooded country like Egypt. The crude gypsum employed by the Egyptians has been analysed by him and proved to have the following constitution: [1]

	Specimen 1	Specimen 2	Specimen 3
Sand	7·6	3·7	2·1
Gypsum . . .	75·4	85·2	89·9
Carbonate of lime	17·0	11·1	8·0
	100·0	100·0	100·0

The material used as mortar is subject to great variation in different examples. Mr. Lucas's analysis of fifteen specimens of mortar from the Sphinx, the Gîza Pyramids, and Karnak gave the following results:

	Max. per cent.	Min. per cent.
Sand	25·5	2·0
Gypsum	89·2	23·4
Carbonate of lime, &c.	71·8	0·7

[1] Lucas, *Mistakes in Chemical Matters frequently made in Archaeology* (*Journal of Egyptian Archaeology*, vol. x, pages 128–31). Other analyses are given in *Ancient Egyptian Materials*, p. 230, by the same author.

As will be shown in Chapter IX, the Egyptians did not lift their blocks by tackle in order to set them. They had to rely entirely on hauling or rolling and the use of levers for this purpose. Thus the mortar played a part of primary importance in Egyptian masonry, since without it large blocks could not have been set at all. In the case of the great blocks already mentioned, which formed the casings of pyramids, considerable experience would be necessary to give the correct thickness to the mortar bed on which the blocks were lowered, so that it would squeeze out into, say, $\frac{1}{50}$ inch when the blocks were finally in position. If pure, the gypsum mortar sets quickly, though this process is very much retarded if it is as impure as that used in ancient Egypt. From tentative experiments made by one of the writers, it seems likely that between the process of lowering the block on to the semi-liquid mortar bed and sliding it 'home', very little time must have elapsed, and this has to be carefully considered when attempting to determine how such blocks were actually laid.

Though mortar on the bed on which a block was to be laid was essential for the laying process, that observed in the rising joints was not. The only useful purpose which it can have served was to fill up any small gaps and render the joint of the same appearance as the bedding-joints, and possibly to prevent plants from sprouting in the joints and thus splitting the edges of the stones. If the mortar were put on the rising joints in any thickness, it would take a very great pressure indeed to force it out to the thickness actually seen in the masonry, a pressure which the Egyptians can have had no means of applying, particularly since the laying of a block had to be carried out during the time the mortar was still viscid (p. 110). Further, it is unlikely that the Egyptians would unnecessarily complicate a process already delicate and difficult unless very little extra labour was involved thereby. The most likely possibility is that the mortar, of suitable consistency, was, as it were, painted on the rising joint immediately before the block was slid into position. Two other possibilities might be imagined, namely, that the mortar was poured into the rising joints after the blocks were in position, or that the mortar found its own way in when the upper course was being laid. It seems very improbable, however, that the mortar could run into such a narrow gap. In some examples of megalithic masonry, such as the south pyramid at Dahshûr and that of Unas at Saqqâra, it seems likely that some pairs of blocks never had mortar between them, though without taking them apart it is very difficult to be certain on this point.

In the Great Pyramid, though the destruction wrought on it is deplor-

able, we are, through it, able to observe how the mortar was made use of
and to study its nature. In the body of the pyramid the blocks are very
roughly shaped and not at all carefully built together. Considerable gaps
are found which are in part filled with a mixture of small pieces of stone in

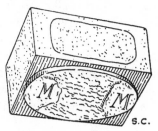

Fig. 73. Lower surface of the
abacus of one of the columns
(no. 46) of the Great Hypostyle Hall at Karnak after
removal for restoration. It
was only mortared at M,M.

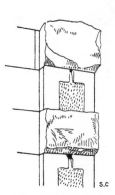

Fig. 74. Roman ma-
sonry at Philae, show-
ing hollowing of joint
to receive the mortar.

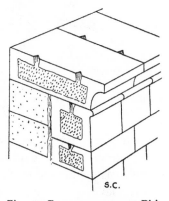

Fig. 75. Roman masonry at Philae, showing hollow rising joints.

a matrix of poor mortar. That this mortar has, or ever had, much cohesive
quality may well be doubted, but the filling in of any considerable holes
between the blocks tends to give increased solidity to the whole mass of
the building.

When the columns in the Great Hypostyle Hall at Karnak fell, in 1899
(p. 75), the stones of these columns came apart and lay on the ground

and partly on each other like rows of huge cheeses. The manner in which
the mortar had been made use of was very clearly revealed. While the
stones lay still untouched by the restorers, one of the writers had an
opportunity of studying them. In certain cases, columns had to be taken
down, for although they had not actually fallen, they had been pushed so
far out from the vertical that the only way to avoid further catastrophe was
to rebuild them. When the abacus of a column (no. 46) was lifted off the
top drum, on which it had rested for some 3,000 years, it could be seen that
the mortar, which had been spread in rather a thick layer, did not lie evenly
over the whole bedding-joint but was concentrated in two places only (Fig.
73, *M*, *M*), the rest of the bed of the abacus not having touched the stone

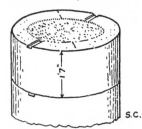 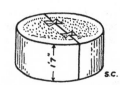 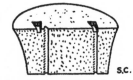

Fig. 76. Roman masonry at Qalabsha showing hollowing of upper surface of the column drums and the outlet-troughs for the superfluous mortar.

Fig. 77. Column drum at Qalabsha. The upper surface has been roughly tooled to give a key to the mortar. Roman date.

Fig. 78. Lower half-drum of a capital at Qalabsha. Roman date.

which supported it. The abacus had to carry the massive monolithic architrave which extended from column to column, and was in turn weighted
down by the heavy roof-slabs which covered the hall. By all theory, the
abacus should have split, but its mass appears to have been sufficient to
withstand the unevenness of pressure to which it had been subjected in
consequence of this economy in mortar.

It is not until Roman times that cement-mortar is found, when an entirely different technique for mortaring the joints appears. At Philae, in
the masonry on the east side of the entrance to the sanctuary, the blocks are
dovetailed and dowelled (p. 112), and under the dowel-hole a groove leads
down to a square recess in the rising joint, possibly so that the mortar could
be run down into the joint whilst the stones were in contact (Fig. 74). In
the cornice-stones which cap the walls in this part of the temple, the channel
between the interior of the joint and the top of the masonry has the form

of two triangular indentations in one of the stones forming the joint (Fig. 75). Though the mortar may have been introduced by the little channels described, it is far more likely that they were escapes for surplus mortar in the joint when the blocks were brought together. At Qalabsha, in the outer court between the pylon and the hall, the tops of the drums of the columns were slightly sunk in the middle and roughened to give a key to the mortar, there being two little troughs on opposite sides of the drum so that the surplus could escape (Fig. 76). By no means could mortar have been introduced through them, which suggests that the channels described above may have been for the same purpose. In the same temple, column-drums without any hollowing of the upper surface can be seen, which are fairly flat and roughly tooled to within a few inches from the edges (Fig. 77). In all these examples, the under surface of the drums seems generally to have been left fairly smooth and not tooled to hold the mortar. At Qalabsha, in some of the half-drums which formed the lower part of the great capitals, two grooves of almost triangular section can be seen running vertically down the joint (Fig. 78). They may possibly have been for the mortar to form a dowel.

It seems that mortar was spread on with the hand, and here and there the impress of the fingers can still be seen sharp and clear. There is no evidence to show that it was ever spread, in Egyptian times, with a trowel, though a plasterer's tool which functions on the same principle is known (Fig. 263).

VIII

HANDLING THE BLOCKS

ONE of the most disputed points in connexion with the mechanical achievements of the ancient Egyptians is the question of what means of handling the blocks they knew, and what they did not know. The view taken by some imaginative writers in the past was that the Egyptians used the most wonderful appliances, and in their publications they even give descriptions of some of them, though their assertions are unsupported by any evidence. On the other hand, certain authors of the last generation were of opinion that not only were the Egyptians ignorant of such means of obtaining great mechanical advantage as the system of pulleys, or the capstan and its derivatives, but that they did not even use the roller or the lever.

If the contemporary records are searched for information on the ancient methods of handling blocks, very little indeed is found. In their tombs, the great architects who were responsible for the building of the temples or the erection of the giant monoliths give practically no information on the means they employed, and classical writers such as Herodotus and Pliny[1] give us such garbled and impossible stories, gathered, no doubt, from ignorant dragomans, that they are of little or no value. Apart from the meagre information obtained from contemporary records, there are two roads by which one can hope to attain to a knowledge of the Egyptians' methods of handling the blocks; one is by a study of such specimens of their appliances as have been found in excavations, and of the traces they have left on the monuments on which they were used; the other is by rigorously ruling out all hypotheses which do not explain every fact which can be observed in the ancient buildings, and by refraining from crediting the Egyptians with any mechanical contrivance unless there is a certain amount of evidence for it, or until every primitive appliance has been proved incapable of carrying out the piece of work which is being studied. Another false piece of reasoning must also be guarded against, which is to assume that identical methods must necessarily have been used in their entirety for handling a colossus and for dealing with small building-blocks.

[1] We have refrained from giving these at length. The more reasonable can be found in ENGELBACH, *The Problem of the Obelisks*, pp. 88 and 91, and an almost complete collection is given in GORRINGE, *Egyptian Obelisks*, p. 154 *et seq.*

One of the most important questions to consider is whether capstans and systems of pulleys were known. As regards the former, though the principle is of the simplest, there is not a particle of evidence that it was ever employed, and every piece of moving work can be explained without it. When, for example, Dhuthotpe moved an alabaster statue of himself, which must have weighed some 60 tons, from the quarry to the Nile (Fig. 79),

Fig. 79. Scene of the transport of a statue in the XVIIIth dynasty.

he did not use a capstan, but mounted it on a sled, running undoubtedly on sleepers, but apparently without rollers, and had it pulled by 172 men,[1] if we take the number of men shown in the actual scene as correct. A man on the toes of the statue is pouring water or oil on the sleepers to lessen the friction, while three others are carrying a large, irregularly shaped piece of wood. It is not at all certain what this is for, but it may be a sleeper which he is bringing from behind to lay in front of the statue. Its irregularity may possibly indicate that it was merely a rough plank, flattened only on its bearing surface. Another inscription relates that it took 3,000 men

[1] NEWBERRY, *El-Bersheh*, i, Pl. XV.

to bring in a sarcophagus-lid, weighing some 18 tons, from the Wady Hammamât to the Nile.[1] With a few capstans, 100 men could have carried out the work in a month. The system of pulleys, and even the simple pulley, must equally have been unknown. No dressed blocks in position show any traces of slots for 'lewises', or other marks which would presumably be present if the Egyptians had used lifting tackle. Traces of recesses, however, for receiving the points of levers are frequent in large Egyptian blocks from the IVth dynasty onwards. In many temples they can be seen in the ends of the roof-slabs and of the architraves. They are

Fig. 80. Transverse section of part of the end of an alabaster sarcophagus-lid, showing holes through which a rope could be passed to assist in getting the lid into place. (XIIth dynasty; Cairo Museum.)

also especially noticeable in the ends of the great blocks forming the lintel of the gateway of Nectanebos II at Karnak. On the lids of large sarcophagi bosses are often left at the end of the lids, which, at first sight, appear to have been for passing ropes round, but the shape of many of them rules out this possibility, since their flatness below and their considerable taper would render them most unsuitable for holding the bight of a rope. It seems that they were for taking the points of levers. The holes in the ends of certain stone sarcophagi of the Middle Kingdom, however, are certainly to enable ropes to be used in lowering or sliding on the lid (Fig. 80). But this is no proof at all that a pulley was used in conjunction with the ropes, since a lever can be used with a rope as shown in the illustration, and this is a common method in modern practice.

In megalithic masonry, such as the casing-blocks of pyramids, small bosses were left low down on the front face of the block for the same purpose. Examples of these can be seen in the unfinished granite facing of the Third Pyramid at Gîza (Figs. 99 & 100), and traces of them can

[1] BREASTED, *Ancient Records*, i, § 448.

Fig. 81. Traces of handling-bosses in the quartzite masonry of
the 'Osireion' of Seti I at Abydos

Fig. 82. Wheeled baggage-waggons, drawn by oxen, used by the Hittites during a campaign
of King Ramesses II; from his temple at Abydos

Fig. 83. Scaling-ladder, fitted with wheels and kept from slipping by a handspike, from the Vth dynasty tomb of Kaemhesit at Saqqâra. This is the only representation of a wheel known in the Old Kingdom. (Photograph supplied by Cecil Firth, Esq.)

also be observed in the partly dressed quartzite masonry in the cenotaph of Seti I at Abydos, popularly called the 'Osireion' (Fig. 81), though here they have been almost removed in the rough dressing of the walls of this building. By no possibility could ropes have been used in connexion with these bosses; further, they indicate that the blocks were handled from the front during the laying process (p. 110). Another argument against the use of tackle for handling the blocks is gained from a study of the models of ships and boats and the representations of them on tomb and temple walls (Chapter V). It is seen that, even in the largest sea-going ships, the halliards which raise the lower boom or yard do not pass through anything like a pulley, but over what appears to be a frame or else through rings attached to the masthead. This lack of means to change the direction of a pull on a rope with any efficiency had a marked effect on the form of rigging used. Had a pulley or system of pulleys been known, it would surely have been used in ships.

Certain recesses in the casing-blocks of the temple of the Second Pyramid have been explained[1] on the assumption that they were to take the ends of great 'tongs', by which they were lifted into place. This, however, does not explain how the casing-blocks of the other monuments in the necropolis at Gîza were laid, since they do not show such recesses. Another suggestion made in connexion with this temple was that the holes observed in the foundations of the pavement were to take the legs of a gyn for raising the statues of the king into a vertical position; again implying the use of tackle. It seems that the holes were, however, to take the feet of scaffold-poles by means of which the statues, &c., were finally dressed (Fig. 232). They would afterwards be covered up by the pavement. Such holes are found all over the foundation recesses of pavements at Gîza; some appear to have been to take the point of levers when used as handspikes, while others may have been used in conjunction with anchorages for ropes.

The pulley involves the use of the wheel, and for heavy work, a very strong wheel indeed. As far as is known, the wheel played a very small part in the life of the ancient Egyptian; the word for it is almost certainly of foreign origin, and it is not found applied to chariots or waggons until the New Kingdom, though this may well be because horses do not appear in Egypt much before that date. The wheels of the known Egyptian chariots are extremely flimsy affairs, and it is doubtful if any wheel built on lines similar to those which have come down to us would take any load or endure hard wear. In the wars of Ramesses II against the Hittites, it is clear, from the

[1] HÖLSCHER, *Das Grabdenkmal des Königs Chephren*, pp. 74 and 75.

camp scenes in the Ramesseum and elsewhere, that he used chariots, but there is no evidence that he used waggons for transporting his gear. The most he seems to have had are quite small carts. The Hittites, on the other hand, had wheeled waggons, drawn by horses and oxen (Fig. 82), piled up with all kinds of goods. In Assyria, solid wheels were used to move heavy weights as early as the eighth century B.C.,[1] but all the known evidence on the methods of transport for building materials used by the Egyptians indicates that the sled alone was used.

It might be considered surprising that no wheeled vehicle except the chariot is represented in the hundreds of scenes on tomb walls which portray the daily life of the dynastic Egyptians, especially since the Nile mud can be made into an ideal surface for wheeled traffic. It must be remembered,

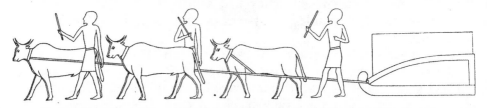

Fig. 84. Block mounted on sled and drawn by a team of oxen, from the Tura Quarries. XVIIIth dynasty. (From DARESSY, *Annales du Service*, xi, p. 263.)

however, that, even to-day, the *fellâh* makes very little use of carts, and many villages in Egypt cannot be approached even on a bicycle. Until quite recently there was no road between successive capitals of provinces in Upper Egypt. The Nile was the highway of Egypt, and the donkey the universal means of transport for light goods over short distances.

It must not be assumed that the use of the wheel apart from chariots was quite unknown, even as far back as the Old Kingdom. In the tomb of Kaemhesit, of the Vth dynasty, at Saqqâra, there is a scene (Fig. 83) of men clambering up a scaling-ladder fitted with solid wheels, which is prevented from slipping out by a man using a baulk of wood as a hand-spike.

To doubt that the lever was used is no longer possible. The recesses in the large blocks and the handling bosses already mentioned could not be for any other purpose than to take the points of levers, and in robbed sarcophagi which are not fitted with handling bosses, it can often be observed that part of the cover has been cut away to enable the end of a lever to be inserted; and, if this is not sufficient, a piece of acacia-wood, with its end

[1] LAYARD, *The Monuments of Nineveh and Babylon*, Pls. X–XVII.

cut to a rough chisel-point, was actually found in a tomb at El-Bersha close to a robbed sarcophagus, which it had obviously been used to open.[1]

The sled mentioned as used for the transport of the 60-ton statue of Dhuthotpe is by no means the only evidence we have of this being the usual method for transporting blocks of all sizes. In the quarries of Tura, there is a scene of a block mounted on a sled being drawn by a team of oxen (Fig. 84). It appears to date to the reign of Amasis I.[2] The obelisks of Hatshepsowet, which the well-known scene in the Deir el-Bahari Temple at Thebes represents loaded together on to a giant barge (Fig. 39), are both on sleds, though here the details of the actual lashing are

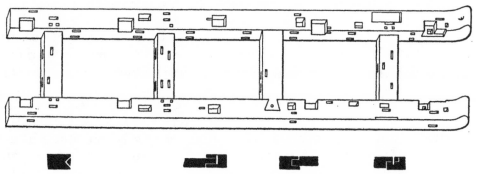

Fig. 85. Wooden sled of the XIIth dynasty from Dahshûr, on which a royal barge had been transported (Fig. 34). The original is 14 feet long. (After REISNER, *Models of Ships and Boats*, p. 89.)

rather vague, since the hauling ropes would never be attached to the middle of the lashings, but rather to the sled itself as in the scene in the tomb of Dhuthotpe, where the lashings, tightened by means of a tourniquet, are shown with great clearness. A large wooden sled (Fig. 85) was found near the south pyramid of Dahshûr,[3] which had been used to transport a royal barge (Fig. 34). Many sarcophagi are also made as if mounted on sled-runners, the outer wooden coffin of Yuya and the mummiform granite sarcophagus of Para'messu in the Cairo Museum being typical examples.

The evidence for rollers having been used in conjunction with sleds is comparatively slight, though their use is almost unquestionable. Examples have been found at Saqqâra (Fig. 267), and in the quarry near the XIIth dynasty pyramid of El-Lahûn, which appear to be of acacia wood.[4]

[1] DARESSY, *Annales du Service*, i, p. 28.
[2] *Ibid.*, xi, p. 263.
[3] The figure is from REISNER, *Catalogue Général du Musée du Caire; Models of Ships and Boats*, p. 89.

Also see DE MORGAN, *Fouilles à Dahchour* (Mars-Juin 1894), p. 83.
[4] PETRIE, *Tools and Weapons*, xlix, nos. 38 and 39.

They are slightly thicker in the middle than at the ends, which are rounded. Similar pieces of wood, which appear to represent rollers, occur in several foundation deposits. The absence of large rollers in the few quarries which have been examined is in no way surprising; then, as to-day, a roller was used until it split too badly to be of further use; it then became firewood. A sound roller was a valuable object, and the chances are much against one being lost. The same remarks apply even more strongly to large levers.

Since Dhuthotpe's statue is represented as mounted on the sled without rollers being shown, writers have believed that they were not used at all. Assuming that the 172 men shown in the scene were the number actually employed, we get a pull of some 8 tons, which might well be sufficient to move it if the sleepers were well greased or watered. It is stated that when the obelisk now in the Place de la Concorde at Paris was being hauled up a very slight incline, mounted on a wooden 'cradle' running on a specially prepared greased way, it took a pull of 94 tons to move it.[1] This would correspond to the force exerted by at least 2,000 men. To pull a colossus of nearly 1,000 tons would have required more than 20,000 men, since the crushing on the runners and sleepers would greatly increase the relative friction factor. Such a number could not conceivably be put on to a block when introducing it into a temple or up an embankment, or loading it on to its barge.

The Assyrians knew the value of the roller as a lessener of friction at least as early as the eighth century B.C. In the scene of the transport of a colossal winged bull[2] at Nineveh, a gang of men are continually supplying rollers under the front of the sled, while others are helping to overcome the initial friction from the back with levers. Between Egypt and North Mesopotamia there were close, though not always amicable, relations from the fifteenth century B.C. onwards, and it is incredible that Assyria should have known the roller and not Egypt. A nation which moved blocks and which never deduced the value of a roller for reducing friction from such homely occurrences as slipping on a walking-stick left on the floor would be sub-human in intellect, which the Egyptians certainly were not. It must be determined, however, why rollers were not invariably used. In the case of moderate sized blocks, and with sufficient men on the spot, the running of the sled over transversely laid sleepers is a considerably quicker process than the use of rollers, which need a good deal of attention to avoid jamming or running sideways. Another disadvantage of the roller is that it requires longitudinal track-baulks, which need careful laying over

[1] LEBAS, *L'Obélisque de Louxor*.   [2] LAYARD, *The Monuments of Nineveh and Babylon*, Pls. X–XVII.

uneven ground, since they have to be supported almost throughout their length.

Evidence has been brought forward that the Egyptians did not use lifting tackle, and all that that implies, but that their only methods of moving a large block were by means of sleds and levers; it remains to be determined how, for instance, a large building such as the Great Hypostyle Hall at Karnak was constructed. It seems that the very simplest method was used, that of filling with earth as the courses grew higher. The first course would be laid, together with the bases of all the columns. The whole interior would be filled with the Nile mud, which hardly requires any ramming, and embankments would also be made all round the *outside* of the course so as to give a platform a few yards wide for handling the blocks from the front. The second course and the corresponding drums of the columns would then be laid and the filling continued. The blocks would be hauled up supply embankments placed at convenient points. The slope of these embankments would depend on the size of the blocks used in the building, and as the building grew higher, would extend far out from it. Using such a method, a maximum number of men could work at the same time, and there is no need to assume any complicated apparatus. It might be considered that to fill such a hall with earth would be laborious beyond the bounds of reason, but this is far from being the case. The hall measures 170 feet by 329 feet, and it can be calculated that the whole of it could be filled, by 1,000 youths, to the depth of the roof of the north and south aisles—that is some fifty feet—in less than six weeks, assuming a six hours' working day and an average carry of 300 yards. To fill to the height of one drum would take less than a week. The modern Egyptians, with their baskets, which they call *muqtaf* or *ghalaq*, can work at the most astonishing pace, and their endurance is phenomenal. The late Director of Works at Karnak, M. George Legrain, used the filling method freely in his repairing work, though he naturally took advantage of modern tackle when it would save time. He used to affirm that he could have carried out all his work without tackle, and using only the primitive methods. To maintain the verticality of the wall during construction would not be a difficult matter, since we know that the blocks were left rough on the faces and dressed after the wall was built. A few pits, lined with brick along the walls, could easily be made, down which a plumb line could be lowered to maintain the general verticality of the masonry. In the case of a batter being required, as in a pylon or mastaba face, a man would be lowered into the pit to adjust the point of the plumb-bob on measuring marks left at the foot of the masonry, and the masonry would be set back to the required batter by calculation from the length of the plumb-line. Egyptian mathematics were well equal to this calculation.

The evidence that the Egyptians made use of embankments of brick and earth is considerable. There is (*a*) written evidence, (*b*) pictorial evidence, and (*c*) remains of at least two embankments in an unfinished piece of masonry. To take these in order; in a papyrus now known as the Papyrus Anastasi I,[1] there is a collection of model letters intended for students to copy. They are supposed to be written by a scribe called Hōri, who twits another called Amenemōpe with not being up to his job (among other failings). In one of his paragraphs, he asks whether Amenemōpe can get

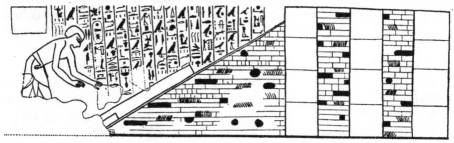

Fig. 86. Scene from a tomb, probably representing the construction of a building by means of a brick embankment. XVIIIth dynasty; Thebes. (From NEWBERRY, *Rekhmara*, Pl. XX.)

out an estimate of the number of bricks required in the construction of an embankment 730 cubits (418·3 yards) in length, 60 cubits (34·4 yards) at the high end and 55 cubits (31·5 yards) broad, having a batter, presumably at its sides, of 15 cubits (8·6 yards). Though the technical terms are extremely obscure, it seems that the embankment was to have been, as it were, a brick box, divided into compartments which were to be filled with earth. There seem to be some errors, probably due to repeated copying, in the figures giving the dimensions of the internal compartments, but the overall dimensions are quite clear.

An even more interesting piece of evidence for constructional embankments is found in the tomb of Rakhmirē' (no. 100) at Thebes, who included many scenes of arts and crafts on the walls of his tomb-chapel. Here can be seen (Fig. 86) an embankment leading up to the top of what must surely be masonry embedded in brickwork. It may possibly represent

---

[1] A translation of the relevant parts of this papyrus may be found in ENGELBACH, *The Problem of the* *Obelisks*, p. 90. The whole papyrus is published by ALAN GARDINER in *Egyptian Hieratic Texts*.

Fig. 87. The unfinished pylon at Karnak (No. 1), showing traces of an ancient constructional embankment on the west face. (From a photograph taken about 1900)

Fig. 88. Remains of the constructional embankment on the east face
of Pylon I at Karnak, showing traces of the internal structure

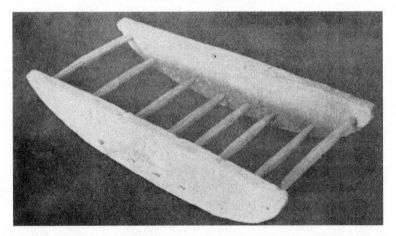

Fig. 89. Ancient model of a wooden appliance for handling blocks,
from a foundation deposit of Queen Hatshepsowet in her temple at El-
Deir el-Bahari. (Cairo Museum)

three columns, and if this is so, the drawing is a truly sectional one. On the other hand, it may be the usual half plan and half elevation so common among the Egyptians (Chapter V) and represent the walls as viewed from the top and the embankment as seen from the side. This, perhaps, is more likely. On the left is seen what may be a roofing-block on the way up, though the writers are at a total loss to explain the details of it. The inscriptions give no descriptive information at all. It is a thousand pities that this unique piece of evidence should have been so ruined by the natives who, until quite recently, made the tomb-chapel their home.

The actual remains of a constructional embankment are still visible on the east and west sides of the great unfinished pylon at Karnak, now known as Pylon I (Figs. 87 and 88). The tremendous masses of masonry forming the pylons appear to have been begun under King Sheshonq I of the XXIInd dynasty. The constructional embankments were apparently made in the following manner: solid and stout walls were built of mud brick, spaced from two to three yards apart, and placed at right angles to the face of the intended pylon. These walls had necessarily to be of some substance, for as the masonry rose, course by course, so did they, until they would be at last of the same height as the pylon, which would have been some 140 feet when finished. As they grew higher, these brick-filled embankments must have been extended backwards so that the slope of the outer face should not become too steep. Between these parallel brick walls, earth—perhaps consolidated with water—was introduced. Here and there also inferior brickwork was built in. The value of brick walls in stiffening the embankment must have been considerable. It is interesting to observe that the brick wall and the filling do not touch the face of the masonry; a gap of about sixteen inches is left. The northern tower of the pylon had a large mass of similar construction remaining in its place, which once partly buried the small temple of Seti II. A great deal of this embankment has now been removed, but it is much to be hoped that, in the interests of archaeology, the mass now to be seen will be allowed to remain; it does no harm and is a leaf of no mean value in the history-book of the ancient Egyptians. In the present ruined condition of the embankments, it is difficult to say which were the supply embankments, and which were merely the platforms in front of the course.[1]

Embankments being well vouched for, the question arises whether the blocks were invariably hauled up the constructional slopes on sleds. It seems that this must very largely have depended on the size of the blocks.

[1] From NEWBERRY, *The Life of Rekhmara*, Pl. XX.

It will be seen, in Chapter IX, that unless we assume that sleds were used at the top of the masonry, it is almost impossible to explain the method of laying the great blocks seen in the Old Kingdom masonry (p. 109). Since the steeper the embankment is, the less material is necessary for it and the less is the space required, it seems very likely that, in the case of moderate sized blocks, they may have been rolled up the slopes, and only mounted on sleds when they had to be moved any considerable distance horizontally.[1]

When a block was being shifted, either on a sled or otherwise, it must frequently have become jammed, and it is interesting to speculate on the extent to which the 'Spanish windlass' was employed in such cases. The principle of this appliance is to pass a rope round and round the object to be hauled and a fixed anchorage and twist the strands between them with a stick in the manner of a tourniquet. Simple forms of this can be seen in the lashings of the statue of Dhuthotpe (Fig. 79), and in the stiffening of the rudder-posts of a ship (Fig. 40). Though there is no proof of it, this principle may have been extensively used by the Egyptians.

In certain foundation deposits of the New Kingdom, small appliances have been found which archaeologists now know as 'rockers' (Fig. 87). Prof. Petrie, in his *Arts and Crafts in Ancient Egypt*, p. 75, suggests that the stone was placed on the appliance and that a wedge-shaped piece of wood was placed below one side of it and the appliance rocked up on to it; another similar piece of wood was then placed on the other side and the stone rocked back in the other direction, this process being repeated by inserting new pieces of wood alternately on each side previous to rocking it, thus gradually raising the block. The appliance may conceivably have been used for this purpose when the block to be raised was so situated that the construction of a ramp was impossible. M. Auguste Choisy, in his *L'Art de Bâtir chez les Egyptiens*, would have us believe that not only were the rockers used in raising the blocks up the sides of pyramids when they were being constructed, but that the brick constructional-embankments (p. 92) were made in the form of huge stairways and that the blocks were raised from one step to the next by rockers.[2] It is far more likely that the rocker, being the ideal appliance for giving small motions with a minimum of effort to anything resting upon it, was used for shifting the block about during the dressing process, and as such was included with the tools in the

---

[1] Another constructional embankment has been traced in the temple of the Second Pyramid at Gîza (Hölscher, *Das Grabdenkmal des Königs Chephren*, p. 72).

[2] The use of rockers for raising the blocks up to the courses in pyramid construction can be shown to be impossible (p. 121).

foundation deposits. The possible use of the rocker in dressing the blocks is discussed on p. 103.

Though the building methods of the Egyptians do not depend on scaffolding, it must not be assumed that this was, to them, an unknown art; on the contrary, there is very good evidence that it was both known and used (Fig. 232) in work on large statues, and it was probably freely employed in facing the masonry of buildings of moderate dimensions, though in the case of very large walls, pylons, and pyramids it is much more likely that this process was carried out to the plane of previously established facing-surfaces (p. 62) while the constructional embankments were being removed.

There is no reason to suppose that the Egyptians were so hidebound by convention that constructional embankments were used for all buildings, however small. Where the blocks could be handled by a party of men without the aid of levers, it is probable that all earthworks, except, perhaps, a steep supply embankment, were dispensed with.

The popular conception of the ancient architects as intellectual supermen has to be considerably modified when an unprejudiced study is made of their works. Amazing as it may seem, no advance was made in their mechanical methods from the IVth dynasty onwards, and it is difficult to determine what was the factor which enabled them to make their early progress. The Egyptian mind was not, in matters unconnected with religion, speculative. His mathematics were so cumbersome[1] as to be inadequate for any really refined calculation, and were rigidly practical. He could use primitive appliances with an almost incredible refinement and was a superb organizer of labour—therein lay his genius. The more, however, his constructional methods are studied, the more one is convinced that if any detail in a piece of work has to be explained by an apparatus of any complication, then that explanation is certainly wrong.

[1] See Chapter XX.

IX

DRESSING AND LAYING THE BLOCKS

AN understanding of the methods used by the Egyptians in dressing and laying stone blocks would seem, at first sight, to be simple enough. In reality, however, the reverse is the case. From classical times down to the present day, the methods used have undergone no very great change. The general principle can be thus outlined: A rough stone block is stood on a flat surface, such as a heavy wooden table (sometimes called a 'banker') and is dressed by means of a chisel or toothed adze ('comb'), and tested for rectangularity by a mason's square. The block is then *lifted* and lowered into place on the course by means of tackle, often in conjunction with some simple holding appliance such as a lewis or tongs. An examination of Egyptian masonry, on the other hand, soon convinces all but the wilfully blind that the underlying principles of the craft were radically different from those mentioned, and that the student must divest himself of all preconceived ideas based on modern practice before he can hope to understand the various processes which, in Egypt, led to a finished stone building.

It is strange that no serious attempt has, to the writer's knowledge, hitherto been made to discuss, in detail, what the Egyptian methods of construction may have been. An inquiry has, therefore, to be conducted from the very beginning. It is for this reason that so much attention has been devoted, in Chapters IV and VIII, to the methods of handling the blocks which were known to the Egyptians, where the evidence brought forward is strongly against their having known or used tackle of any kind for lifting heavy weights. Since the Egyptians have left practically no information at all on their building methods, the student has to rely on deduction based on evidence in the monuments themselves, and, since an assertion, or even a suggestion, that this or that method was or may have been employed is utterly valueless without the evidence on which it is based, it has been felt necessary to give that evidence as fully as possible.

Egyptian masonry can be divided into two classes, namely small-block masonry, where the blocks were sufficiently light to permit them to be lifted by a party of men, and megalithic masonry, where the builders, being unacquainted with lifting tackle such as the system of pulleys, had to roll or slide the blocks into position.

Given a sufficiency of binding material, once the idea had been originated

Fig. 90. Masonry in the temple of King Zoser, north-west of the Step Pyramid; IIIrd dynasty, Saqqâra. (Photograph by the Antiquities Department, Egyptian Government)

Fig. 91. Panelled boundary-wall of the Step Pyramid at Saqqâra. IIIrd dynasty

Fig. 92. Top of one of the bays (*Fig.* 91, A) in the boundary wall of the Step Pyramid at Saqqâra, showing the poorness of jointing in the small-block masonry of King Zoser

Fig. 93. Ribbed columnoid in the cross hall at the west end of Zoser's colonnade at Saqqâra. The courses are not parallel

of using stone as a medium for building, two types of masonry would tend to develop; the first method would be to build with the flattest face of the block to the front, and the second would be to endeavour to obtain as close a fit as possible between neighbouring blocks. The first was used freely at most epochs for retaining walls, but it must have been from the second method that the fine masonry in Egypt took its rise, where the object may have been to imitate the fine and regular joints observed in good brickwork. It might be imagined that the next step would be to cut the stone into rectangular blocks of equal depth and to lay them just as blocks are laid to-day. But a brick is not cut; it is cast, and the Egyptian does not seem to have acquired the idea of testing a block with a square until it was perfectly true. This may possibly be partly accounted for by the fact that he never learned, even in the smaller brick walls, the art of internal bonding (p. 113). Hence the only joints which he would aim at making close would be the joints on which the blocks lay, which are known as the *bedding joints*, and those between neighbouring blocks along the course, or *rising joints*. It is probable that it never occurred to the Egyptian that there would be any advantage in using truly rectangular blocks; at any rate they can almost be said not to exist until the very latest times.

The masonry of Zoser (Figs. 90–5) is inferior to the better examples of later times in that the fineness of the joints between two adjacent blocks, which appears so good when viewed from the front, only extends inwards for at most a couple of inches; afterwards the joints become wide and irregular and are filled in with thick white gypsum mortar (Figs. 92 & 94). The tops of the blocks were generally made to slope downwards from the face, probably after they were laid (p. 107), thus enabling the bedding joints to be made close at the face of the wall, since the larger the surfaces are which are required to be in close contact, the greater must be the pressure exerted in bringing them together in order to squeeze out the superfluous mortar. The difficulty of obtaining a close joint between contiguous blocks was admirably overcome in the succeeding dynasties, where the rising joints between blocks of the largest size are close from front to back (Fig. 96). In the Zoser masonry, fineness of jointing at the face of the walls was only obtained at the expense of solidity.

Where the masonry was to be covered by a pavement, as in the casing-blocks of the Step Pyramid (Fig. 95) and many of the shrines round it, the front of the blocks was left in the state in which they were laid, and it can thus be deduced that it was after the laying had taken place that the face of the wall was dressed smooth. This is confirmed by the fact that at the internal corners of walls the rising joints do not come at the corner of the wall itself. This feature is noticeable throughout the dynasties (Fig. 230). It seems also that in the Zoser masonry the tops of the blocks were likewise left in the rough until after they were laid, since in the columns of the cross-hall of the western colonnade (Fig. 94) the bedding joints, though flat, are not parallel, which could hardly happen if the blocks were laid with their top surfaces dressed. This peculiarity is not confined to the IIIrd dynasty.

The manner in which a block was laid in the Zoser and other small-block masonry appears to have been as follows. It was brought up on to the course and laid close to that which was last laid, and two surfaces, extending inwards from the front of the blocks for about two inches, were

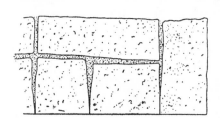 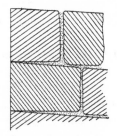

Fig. 94. Plan and section of an example of Zoser's small-block masonry at Saqqâra. (Slightly exaggerated.)

made to fit. The faces of the blocks behind these dressed surfaces were cut away so that there should be no contact (Fig. 94). Since a block could be lifted, it could be tried against its neighbour as many times as necessary until the required closeness of fit was obtained. The rising joints were made, in general, approximately vertical, but if the conformation of the blocks was such that an oblique joint was more convenient, then an oblique joint was made. In what may be the earliest example of the small-block masonry, namely the festival temple, the traces of laborious fitting are more frequent than in the remainder of the buildings round the Step Pyramid. Here it can be seen that the bedding joints are at times very irregular. It may be that in this example, the part of the top of the course on which a block was to be laid was dressed immediately before this was done. The tops of the courses are not smooth, but are usually scored with the marks of tools. This may have been occasioned partly by the dressing of the rising joints when the blocks were standing on the course, but may also have been for the purpose of giving a key to the mortar.

Fig. 95. Remains of the casing-blocks of the Step Pyramid of Saqqâra (IIIrd dynasty). The dressing of the rough faces of the blocks, after they were laid, has not been completed on the lowest course, probably since it lay below ground or pavement level. (Photo. by Cecil Firth, Esq., Antiquities Department, Saqqâra)

Fig. 96. Casing-blocks on the north side of the Great Pyramid. The joints in these blocks do not gap more than $\frac{1}{50}''$ at any point in their surfaces. Behind are the 'packing-blocks' of the second course

Fig. 97. South face of a mastaba at Gîza. The levelling of the casing-blocks, which have now disappeared, was carried, at AB and CD, a short way into the core-blocks. At EF and GH, traces of the levelling of the intermediate casing-blocks can be seen. The core-blocks are about four feet high

Fig. 99. Granite casing-blocks, only partly faced, near the entrance to the Third Pyramid a Gîza, showing bosses under which the points of levers engaged during the laying. The averag height of a course is about four feet

Fig. 100. Granite casing-blocks on the east side of the Third Pyramid at Gîza, where the facir has been nearly completed, showing oblique rising joints

A peculiarity in the small-block masonry of Zoser is the number of patches which are noticeable at the joints. It seems likely that, during the adjusting or the dressing process, the blocks frequently became chipped, and that when this happened, the faulty parts were cut out, oblong slots being formed, the sides of which taper inwards, and pieces of stone were cut to fit into them. The patching was done with considerable neatness, and occurs most frequently in the festival temple and least in the cross-hall

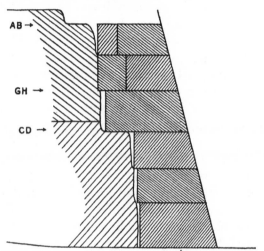

Fig. 98. Restoration of section of part of the casing- and core-blocks of the mastaba shown in Fig. 97.

(Fig. 3). The correction of faults by patching occurs throughout the dynasties, but not so frequently as in Zoser's masonry.

The methods used by the IIIrd dynasty architects did not die out during the ensuing dynasties. As long as a block could be conveniently lifted, they continued to be used in all but the finest masonry. In many of the IVth and Vth-dynasty mastabas at Gîza, masonry can be observed where the blocks fit tightly at the face but the joint only extends inwards for at most a couple of inches. When, however, it was desired to lay blocks, especially those of large size, with a perfectly fitting rising joint from the front to the back, a considerable modification of technique had to be evolved.

Before entering into the question of the manner in which the blocks in fine megalithic masonry were dressed, it is necessary to determine how many faces were made smooth before a block was laid. It has been shown that in the Zoser masonry the likelihood is that the only faces cut before laying were those which were to form the rising and the bedding joints. In the megalithic masonry the same practice appears to have been followed. From the many unfinished pieces of masonry which have been preserved (Figs. 99 & 100)[1] it is certain that the front of laid masonry was dressed after all the blocks forming the wall or building had been laid. The evidence that the tops of the blocks were dressed after they had been laid is almost conclusive. In certain of the Gîza mastabas from which the packing- and the casing-blocks have disappeared, it can be seen (Figs. 97 & 98) that the flattening of the top of a course had been carried back to a distance sufficient to involve part of the core-blocks as well. This can also be observed in the Great Pyramid, where the tops of some of the packing-blocks (p. 122) have only been partly levelled, the difference between the levelled surface and the original top of the block being a matter of several inches. Another indication is that in many examples of masonry, the bedding joints of one course, though flat, are sometimes not parallel with those of another (Fig. 120), which would only happen if the tops of a series of blocks were dressed together after laying. Further, it will be seen (p. 107) that in the form of masonry in which blocks of unequal heights were employed, it is only by such an assumption that a satisfactory explanation can be found for its peculiarities.

The backs of the blocks in a building were, in most cases, only very roughly dressed, if at all. Exceptions are, however, encountered, of which the finest examples are the casing-blocks in the Great Pyramid (Fig. 96). The fact that the Egyptians could, if they so wished, obtain quite a good fit at the back of their blocks as well as at the sides is important in the consideration of the methods used in dressing the blocks.

Fine megalithic masonry shows two striking peculiarities to which modern masonry furnishes no parallel. First, the blocks, though they may fit perfectly against their neighbours from the front to the back, are not truly rectangular in any sense and often not even approximately so, thus giving rise to what may be termed 'oblique' rising joints (Figs. 99–101, 110 & 111); secondly, in many buildings the blocks used in a course are of very unequal heights (Figs. 101 & 102); sometimes both features occur in the same piece of masonry (Figs. 101 & 103).

The presence of oblique joints between successive blocks certainly rules out the mason's square as having played any part in their shaping, though squares were known to the Egyptians (Fig. 264), and must have been freely used in the facing of the angles of laid masonry. Oblique joints

[1] See also Figs. 81 and 234.

Fig. 101. Wall of a Vth dynasty mastaba at Saqqâra, showing oblique joints. (Photograph supplied by Cecil Firth, Esq.)

Fig. 102. Masonry with blocks of unequal heights. Temple of Amenophis III
at Luxor, east side

Fig. 103. Wall of Tuthmosis III, sou
of the sanctuary at Karnak

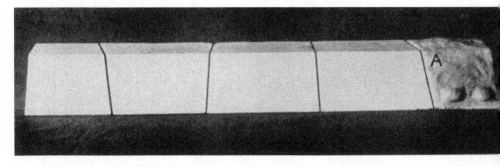

Fig. 104. Model, illustrating masonry of 'Type A', on the front face of which a batter has bee
cut, disclosing the obliquity of the rising joints

appear, at first sight, to be entirely without motive, but their presence is so common at all periods that a mechanical reason has to be found for them, since it cannot be assumed that they were made for fun or as a *tour de force*.

If the masonry containing oblique joints be carefully examined, it will be noticed that, very often, in a course, the planes of a series of rising joints are all approximately at right angles to a plane, and that this plane is either the bedding joint (Figs. 104 & 105) or the vertical plane through the front edge of the course (Fig. 106). More rarely, rising joints are encountered

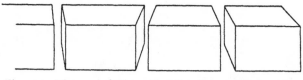

Fig. 105. Masonry of 'Type A'. The planes of all the rising
joints are vertical.

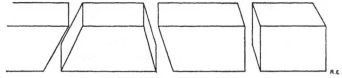

Fig. 106. Masonry of 'Type B', where the planes of all the rising
joints are at right angles to the front faces of the blocks.

where the planes are oblique in every sense. These forms demand close study.

The fact that the planes of the rising joints are oblique shows that one block was shaped to fit another particular block; that they are often closely at right angles to a plane suggests that the blocks were lined up in some manner for dressing and that parallel, and often vertical, planes were cut between each pair of blocks, and that the bedding joints (upon which they would eventually rest) were formed by reducing the top or a side of the whole line to one vertical plane or else by making the top of each block level, the latter method being far less probable. A method by which the rising joints might well have been formed is by measuring off equidistant pairs of marks at the end of each pair of blocks (Fig. 107; $ab = cd$; $a'b' = c'd'$) and by dressing the faces of the blocks to vertical planes below the two points ac, bd, $a'c'$, &c. The presence of a rising joint oblique in all senses could be explained by assuming that, instead of dressing the block below two points to a vertical plane, three equidistant pairs of marks were measured off on each pair of blocks, and that the rising joints were dressed to the plane of the three marks on each block. If such work were done with great accuracy, the blocks would fit perfectly together when laid on their bedding surfaces without there being any need for preliminary trying of one against the other. It is more than doubtful, however, if two parallel planes could be cut accurately enough to ensure the fine fit observed in so many of the ancient monuments; indeed, competent stonemasons who have been interrogated on this point have been unanimously of opinion that it is not possible, though perfect flatness and verticality can be obtained without

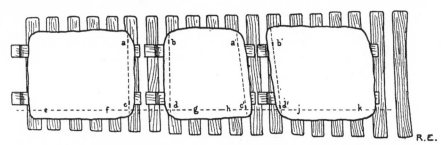

Fig. 107. Plan of blocks lined up on their sleds for dressing, to illustrate the likelihood that the planes between successive blocks were made parallel.

any great difficulty; but the fact that the rising joints are only approximately at right angles to a plane[1] makes it impossible that they can have been so dressed and laid without any previous fitting. If, therefore, it is assumed that the blocks were tried one against the other before laying, the dressing process becomes simpler, and can be thus formulated: The blocks having been lined up in the manner already described, flat surfaces are cut at the ends of each of them, often vertically below two marks (Fig. 107), and these surfaces are rendered perfectly flat and *as nearly parallel as possible*. These flat surfaces are then brought into close contact, and one face of the whole line of blocks reduced to one flat surface to form the bedding joint. If this explanation be true, the method has to be determined by which large blocks can be brought into close contact. If the blocks lay on sleds (p. 89) without projecting runners, this would be by no means impossible, though the sled is not the most handy appliance for giving small motions to a block. The appliance which has been called a 'rocker' (Fig. 89) is, on the other hand, ideal for the purpose, since on it

[1] An ingenious and accurate apparatus for measuring the angle of a plane with the vertical was designed and supplied by Dr. H. E. Hurst, Director of the Physical Department, Egyptian Government.

a block can easily be tilted forwards and back and turned and raised in the manner already described by inserting large wedges (Fig. 267) below the 'runners'. Since the lower surface of a rocker is, in the ancient models, slightly flattened, a block can also be dragged along on it, though for this it is inferior to the sled.

In the illustrations (Figs. 108 & 109)[1] an attempt has been made to represent part of the suggested dressing process by means of models. The blocks are lined up on the rockers, which lie on sleepers embedded in a chip heap. Approximately parallel but perfectly flat surfaces have been cut between each pair[2] and the blocks have been brought into close contact. The tops of all the blocks have then (Fig. 109) been made flat and level[3] to form a bedding joint. An alternative method would have been to reduce the front face of the whole series to a plane, which in practice would probably have been a vertical plane.

It will be observed that, provided the blocks are laid on a flat bed, they will fit with an accuracy depending on the care with which their dressed surfaces have been brought into contact. In practice, as soon as the foreman was satisfied that the two blocks lay tightly together, the rocker would probably be partly embedded in chips to prevent any further motion.

The method described can be extended, in theory, to any number of blocks, for if the planes between each pair had been cut very nearly parallel, the small errors could be corrected by adjusting the blocks vertically and horizontally when they were brought into contact.[4] Other considerations, however, tend to indicate that there must have been a very definite limit to the number dressed together (p. 111).

In the case of the immense blocks, such as are seen, for instance, in the granite casing of the IIIrd Pyramid at Gîza (Fig. 99) and on the limestone casing of the Pyramid of Unas at Saqqâra (Fig. 111), it might be doubted whether any rocker could be constructed to take their weight. It has been remarked, however, that bringing them into contact on their transport sleds is not an impossibility, though it would admittedly take a much longer time.

Masonry of 'Type A' and 'Type B' (Figs. 105 & 106) is contemporary in the Old Kingdom, but after that time it appears that type B is

---

[1] In the illustration two of the joints have been represented as having been dressed to a vertical plane, while the third is oblique in every sense.
[2] Either when on the rockers or while still on the sleds on which they were transported from the quarry.

[3] For notes on flattening and levelling a surface, see p. 105.
[4] It will be observed that the more accurately the planes forming the rising joints are made parallel, the less will be the waste of stone when the bedding joints are cut.

exclusively employed. Occasionally both types occur in the same wall. In the remains of the wall of the funerary temple of the Pyramid of Unas, shown in the illustration (Fig. 110), the angle ABC is a right angle and so is the angle DCE, though the planes of the rising joints are oblique. In the next joint to the left on the bottom course, not shown in the photograph, the plane of the rising joint is oblique in every sense. Such examples of the different forms of jointing in one wall are, as far as the writer is yet aware, rare.

An advantage of 'Type A' masonry, where the tops of the line of blocks were flattened or levelled and where the blocks were laid upside down to the position in which they were dressed, was that, if necessary, another line of blocks, which might form, for instance, packing-blocks in the case of a pyramid, could be dressed at the same time, and made to fit against the first line. If the packing-blocks on the north side of the Great Pyramid be examined it will be observed that the fit between them and the casing-blocks (Fig. 112) is good, though not nearly as perfect as that between two casing-blocks, and it will also be noticed that the rising joints are very closely vertical, though oblique to the vertical plane through the pyramid edge. Further, each packing-block, generally speaking, lies behind a casing-block, which could hardly happen fortuitously, and which gives good reason to suppose that not only were the casing-blocks lined up in the manner described, but that the corresponding packing-blocks were lined up alongside, and that parallel vertical plane surfaces were also cut between the casing- and the packing-blocks and between successive packing-blocks. When, after a series of blocks had been dressed, they were brought together so that the tops of the blocks could be cut to form the bedding joint (Fig. 109) it must often have happened that the packing-block did not lie exactly behind its casing-block. A piece had therefore to be cut from the corner of one or the other before they could be brought into contact. Such cuts can be frequently seen, and they are generally quite roughly carried out. The position of the packing-blocks (and therefore presumably of the casing-blocks) was marked, by means of a saw-cut, on the flat bed on which they would lie. This may have been done by measurement before a series of blocks was laid in order to ensure that the rising joints would break with those of the course below.

Though the method by which it has been suggested that the blocks were dressed may seem fantastic in the extreme, it is not necessarily so. First, it is the only explanation yet brought forward which accounts for the curious features observed in Egyptian masonry; secondly, it involves the minimum

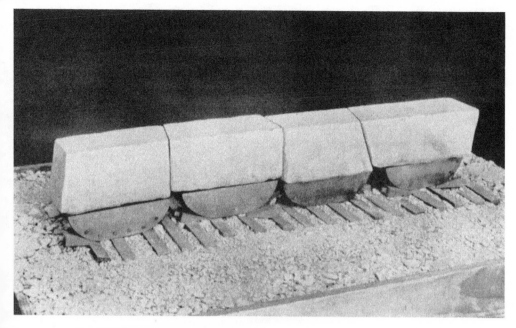

Fig. 108. Models of blocks, between each pair of which approximately parallel faces have been cut, brought into close contact

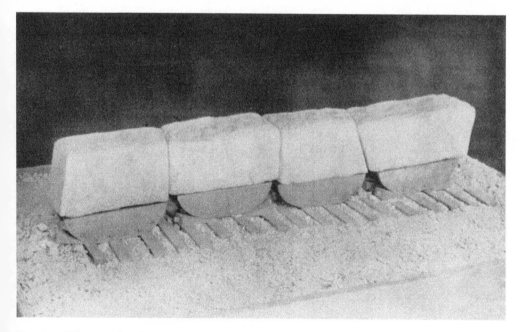

Fig. 109. The same blocks, after the tops of the series have been made flat to form a bedding joint

Fig. 110. Masonry in the wall of the funerary temple of the Vth dynasty pyramid of Unas at Saqqâra. The angle ABC is a right angle, but the plane of the joint AB is not at right angles to the face of the wall. Similarly though the angle DCE is also a right angle, the plane of the joint below is at a slant to the bedding joint. (Photograph supplied by Cecil Firth, Esq.)

Fig. 111. Remains of the casing at the entrance to the pyramid of Unas at Saqqâra. The part A–B, though now broken, consisted of one block. (Photograph supplied by Cecil Firth, Esq.)

amount of handling of a block before laying; thirdly, a maximum number of men can be employed simultaneously without their unduly interfering with one another. If this be the true explanation, it is but another of the many examples of the magnificent powers of organization possessed by the ancient directors of works.

The actual method of obtaining a vertical plane below two given marks, which seems to have been found most convenient when dressing a block, cannot be detailed very precisely, though valuable hints can be found in

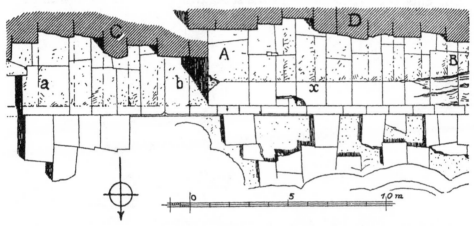

Fig. 112. Plan of part of the north face of the Great Pyramid. AB, casing-blocks (see Fig. 96); c, packing-blocks for first course of casing (also shown behind AB; D, remains of packing-blocks for the second course. ab, paving-blocks on which the casing of the pyramid is laid. (After BORCHARDT, *Längen und Richtungen der vier Grundkanten der grossen Pyramide bei Gise*, Plate 3.)

tomb scenes. In the tomb of Rakhmirē' at Thebes (no. 100) there is a scene of masons dressing the face of a block of stone. Two stages of the work are shown, but unfortunately not the initial stage. On the left (Fig. 113), two workmen are depicted, each holding a short piece of wood, both pieces being of equal length, whose tops are connected by a string. An ancient set of these 'boning-rods' is preserved in the Cairo Museum (Fig. 265). One of the workmen is indicating to another how much material he is to take off, which he has gauged by standing a third piece of wood, of equal length to the other two, up against the taut string. It must be remembered that, in the scene, the boning-rods are represented as flat against the face of the stone, whereas in reality only their tops would be visible. The boning-rods are obviously standing on surfaces to the plane of which it was required to reduce the whole face. In the right scene, the block has been flattened as far as could be done by the previous method, and the final stage in the dressing is being carried out by a man with a chisel, which he is tapping with his fist instead of with a mallet, his work being tested by his companion by means of a stretched thread held against the face of the block. It is hardly necessary again to remark that, to define a required plane surface, three points, which may be termed 'facing surfaces' (p. 62), are necessary unless the plane is to be vertical, in which case only two are needed. In the scene it appears that the workmen are

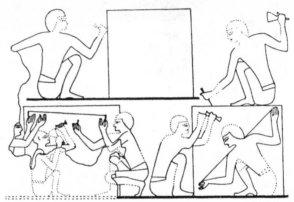

Fig. 113. Ancient scene of workmen dressing a stone block. Below, on the left, three men are flattening the surface by means of 'boning-rods', while on the right the last touches are being given, and the flatness tested by holding a thread against the face. (From NEWBERRY, *Rekhmara*, Pl. XX.)

reducing the face of the block to a vertical plane, but no clue is given as to the manner in which the facing surfaces were obtained. It may well have been some simple combination of the plumb-line and the boning-rods. If the blocks were on the rockers during the dressing process, the final testing for verticality could be performed by means of a plumb-line held directly against the face of the block.

A problem which demands consideration is *why* the Egyptians so often laid their masonry with oblique joints. It might be said that they did so because there was no particular need for them to do otherwise if the blocks were required to be laid in a certain order. This, however, hardly answers the question, since it cannot have escaped the Egyptians that there would be an advantage in having the planes of *all* the rising joints parallel to one another, so that, in the event of any modification of the order of laying,

Fig. 114. After the first course is laid, marks are made on the blocks to indicate where blocks 4, 5, 6 of the second course will lie

Fig. 115. The beds for the blocks of the second course are prepared. The amount of stone cut away depending on the height of blocks 4, 5, 6

Fig. 116. The second course is laid, and marked to show positions of blocks 7, 8, and 9

Fig. 117. The beds for blocks 7, 8, and 9 are prepared

Fig. 118. The third course is laid   Fig. 119. The front face of the wall is dressed

Figs. 114–19. Models illustrating the ancient manner of constructing a wall with blocks of unequal heights

such as one block proving defective or becoming broken in the handling, any block could be laid against any other. It is likely that the idea underlying the use of oblique joints may have been that of cutting the stone as little as possible. This must especially have applied to the great granite blocks which formed the lower sixteen courses of the Third Pyramid at Gîza. The method of dressing granite blocks by pounding with balls of dolerite was most laborious (p. 27); so much so, in fact, that this casing was never completed but left in the half-and-half stage (Figs. 99 & 100). In dressing such blocks, if advantage could be taken of the irregularities of any pair by making an oblique joint between them, this would certainly be done.

Though in many walls and other buildings, the blocks were delivered from the quarry so nearly uniform in height that, when the tops of the courses were dressed, they could all be reduced to one plane, this was not always the case. In many buildings (Figs. 101 & 102) blocks of very different heights were used. If such blocks had been dressed completely and laid as they were, the result would have been an inordinate number of straight joints and a consequent lack of solidity in the completed structure. To obviate this, the Egyptian marked out, on the rough top of the course already laid, the place where the ends of the blocks which were to be laid on it would come, the marking often being done with a saw-cut. The spaces between the cuts were then levelled to any convenient depth considered necessary, and the next course of blocks laid on the beds thus prepared for it. Though extreme examples of this masonry are often seen (Fig. 82), the rule seems to have been to attempt to keep the beds of a certain number of successive blocks at one level. The probable process can be illustrated more clearly by means of models (Figs. 114–19), where the blocks waiting to be laid are lying on their sides with their bedding joint to the front. This is very likely how they lay in actual practice, and they may well have been dressed in this position. It is possible that, in a wall, the blocks were arranged on the ground at least two courses ahead of that actually being laid. Though a course has been described as having been levelled, in some cases the beds are not level, but only flat. This peculiarity, well shown in the little temple of Amenophis III at El-Kâb (Fig. 120, *AB*), naturally tends to make a poor joint between the last block on the slanting bed and that on a bed that is level or at another angle. The presence of such slanting beds is one of the most convincing proofs that the top of a course was dressed after laying.

Though the masonry with blocks of unequal heights appears in the great

majority of cases to be of 'Type B' (Fig. 106), in some of the granite
walls of the valley temple of the Second Pyramid not only can rising
joints which are oblique in every sense be observed, but also bedding joints
which follow almost zigzag courses. Though it is perhaps mechanically
possible that the blocks were dressed in the manner suggested for casing-

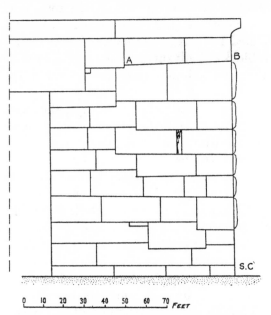

Fig. 120. Elevation of half the façade of the small
desert temple of Amenophis III at El-Kâb. This
temple has been constructed of blocks of unequal
heights. It will be noticed that the bedding joints
of some of the blocks (A B) are at a distinct slope.

and packing-blocks of pyramids (p. 122), and laid as a wall instead of as a
course, it is hardly likely that this was done, since it would mean that the
tops of the blocks were dressed before they were laid, which, as has been
shown, is contrary to the practice of the ancient builders, who were hide-
bound by convention and worked by rule of thumb. It is more likely that
the blocks were taken up to the course and laboriously fitted against their
neighbours, though measurement may have played a part in the process.
It must be remembered that the granite was obtained, not from a smooth
quarry face as in the case of the soft rocks, but from boulders, which would
result in its being supplied in a diversity of shapes. It seems that the builders

found it more convenient to expend time in fitting irregular blocks together if by this means they could in any way reduce the amount of stone which had to be removed from the blocks by the very slow process of pounding with balls of dolerite (p. 27). The joints in the granite walls of the valley temple of the Second Pyramid are not comparable in fineness with those of the casing-blocks of the pyramids.

The question of the dressing of blocks in normal Egyptian masonry having been discussed in some detail, the next problem is how they were actually laid. Though walls, column-drums, architraves, and roof-beams all offer their own special difficulties, a consideration of them all is outside the scope of the present volume. One instance must suffice, and if one of the most difficult cases is chosen, there need be little fear that an explanation of the method of laying the others will not be forthcoming.

The casing-blocks of the Great Pyramid (Fig. 96) are about the best examples which could be taken, since they show finer joints than any other masonry in Egypt, and perhaps in the world; they are nearly as finely jointed to the packing-blocks behind them, and they are of great size, some being nearly fifteen tons in weight.

Several factors have to be taken into account before attempting to determine the process of laying such blocks.[1] It has been shown that the sole means of removing them from the sleds was by levers, and that no lifting tackle was used (Chapter VIII); in considering the use of levers, it has to be borne in mind that very little forward motion can have been obtained with them. Though here and there holes can be seen in the pavement or on the courses in which levers may have engaged when used as handspikes, or in which the fulcra of the levers may have been anchored, they must be disregarded in the inquiry, since they are exceptional. For instance, in the blocks of the Great Pyramid, though the pavement extends for about two feet in front of the blocks, there is no trace of anything which might have served as an anchorage, neither is there, on the top of the course, any apparent means by which a block might have been eased *along* the course to bring it close up to its neighbour. The most that can be said regarding the occasional holes in the pavements and courses is that they may have been brought into use if a block proved refractory—if something went wrong.

Let it be assumed that it is required to lay a block for the casing of a

---

[1] The reader's attention is again called to the function of mortar in megalithic masonry. It is, as it were, a lubricant, on which the block could be slid into position before it dried, and is in no wise a cement (see p. 80).

pyramid, weighing fifteen tons, whose top and front face are rough, the latter having handling bosses under which the points of levers can engage, and that this block, by careful dressing (p. 103), will, if properly pushed home, fit both against its neighbour along the course and against the packing-block behind (p. 104). Let it be also assumed that the top of the course on which it is to be laid is dead flat and that, in front of the course, there is a platform of brick and rubble (p. 92) extending some forty feet out from the stonework. Such are the conditions under which a block was apparently laid in the casing of the Great Pyramid. The block would probably be brought[1] along the course (if it was not the lowest) on its sled[1] until it lay within a couple of inches of the blocks against which it was to be laid. The next problem is the removal of the sled, which would not be simple. Almost the only method, in the absence of lifting tackle, would be to raise the block slightly by jamming it by means of levers, acting from the front under the bosses, against the packing-blocks behind. The sled could then be taken away, and the block let down on to some packing at a lower level. Since levers acting on fulcra which are not anchored can only give a forward motion for a very short distance, it seems likely that the lowering would have to be done in five or six stages. When at length the block was resting on the packing only an inch or two above the bed it would be ready for the mortar. The problem of getting the block very tightly against the next casing-block has, however, not yet been considered. The rising joints of such blocks which can be studied in this and other pyramids show no trace of levers having been introduced beneath their edges, nor against the faces. Had they been eased home by means of levers, traces would surely have been left. A method of great simplicity would be to put a baulk of timber across the free corner of the outer rising-joint face and let a gang of men haul it home by ropes attached to the two projecting ends of the baulk. It seems the only method which would give the required result, though there is no proof that it was used. Before releasing the packing below the block, the mortar would be prepared so as to be of suitable consistency. What this may have been, experience alone could tell. It suffices to say that it was such that the weight of the block, while it was being slid into position on it, squeezed it out into a film not more than $\frac{1}{50}$ inch thick. It is possible that the mortar, in an almost liquid condition, was slopped under the block and smeared lightly over the vertical joints while the block was still jammed against the packing-blocks behind it after

---

[1] On to which it must have been turned from the rocker (p. 94). It must be noted that the sled would have to be without projecting runners.

the last packing had been removed, and that it was let down on to its bed and hauled in against its neighbour immediately afterwards. The reason why it is likely that the last operation was done with speed is that mortar, under the conditions stated, would only remain viscid for a short while, after which the block would no longer be movable. It would be only in the last stage that the men on the levers might attempt to get as much forward motion as they could, in order to make the block lie as tightly as possible against the blocks behind it.

On the assumption that the theories just outlined are, in the main, correct, the question arises whether the whole course was laid from one end to the other or whether the laying was carried out from each end simultaneously and a block inserted to fill the gap in the middle. Though no certain answer is forthcoming, it seems safe to assume that, at times, an accident happened to a block during the dressing or transport, which rendered it unfit for laying. One therefore has to believe that the Egyptians could, if they so wished, cut a block to fit between the oblique rising joints of two already laid. This would happen regularly if the laying were done from two or more points on the course. It is hardly safe to assume that the occasional presence of relatively small blocks in the casing of the south pyramid of Dahshûr and of the Second and Third Pyramids of Gîza resulted from the causes stated, but a pair of small blocks in a course of the Third Pyramid (Fig. 100) gives a strong impression of having been inserted after those on either side of them were in place.

Since the Egyptians must have been able to fit a block against one already laid, it might well be asked whether, in the case of the casing-blocks of the Great Pyramid and of other fine megalithic masonry, it was not the regular practice to bring the block up on to the course and dress its rising joint parallel to that of the last block laid, as appears to have been done in the Zoser masonry (p. 98). Two reasons may be advanced against this having been the case. First, on the tops of the courses in the best megalithic masonry, no traces are found which might indicate that tone dressing had taken place there, and traces would certainly be expected similar to those seen on the pavement of the Great Pyramid, which resulted from the dressing of the front face of the lowest course of casing-blocks which lay on it. Here the position of the edge of the pyramid can be determined at many points where the casing has disappeared. Secondly, fitting one block to another laid block, when it had to be done, must have been a very lengthy process, whatever appliances were used. To carry it out hundreds of thousands of times for the casing of a pyramid would take time beyond all reckoning. The fact that the great pyramids were each constructed during a king's reign is difficult enough to explain, even assuming a comparatively rapid process to have been used by means of which many blocks could be dressed simultaneously.

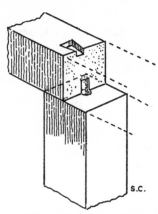

Fig. 121. Dovetail recess in the architraves of 'The Temple of the Sphinx' at Gîza, with hole for dowel passing down into the top of the column.

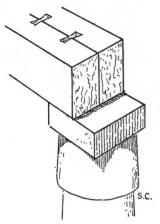

Fig. 122. Double architrave united by dovetails. Medînet Habu.

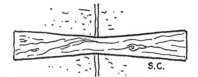

Fig. 123. Dovetail and wooden cramp, nearly five feet in length, in an architrave in the Court of Sheshonq at Karnak.

The use of dovetails to connect two blocks is known from the Old Kingdom to late times, but they are not by any means always used, even in the best masonry. In the Great Pyramid they are not found, nor in the majority of the finest mastabas. In the temple called 'The Temple of the Sphinx' they occur between the ends of the architraves (Fig. 121), where they were also often accompanied by peg-dowels passing down into the tops of the columns. At all periods, they are very commonly used between the blocks of a double architrave (Fig. 122). In some buildings they are of very large

size, those uniting the architraves in the Court of Sheshonq I at Karnak
being nearly five feet in length (Fig. 123).

Dovetails have been found in wood, in lead, in copper, and occasionally
in stone. Each variety can be seen in the Cairo Museum. Their function
is undoubtedly not that of strengthening the masonry. The tensile strength
or the capacity to resist shear in, for example, a wooden dovetail would be
quite insufficient to prevent any settling or bulging tendency in a wall.
The most that can be said is that they were to keep one block against
another while the mortar was drying—especially if the beds on which they
were laid were not very level—and perhaps, in masonry of moderate-sized
blocks, they may have played a part in preventing the blocks from becoming shifted while the other blocks near them were being handled. In

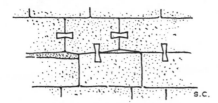

Fig. 124. Horizontal section of a wall
of Ramesses III at Medînet Habu.

many examples of masonry which have had to be taken down for restoration, the dovetail recesses have been found to be filled with mortar. It is
likely, once the mortar had set and the neighbouring blocks had been laid,
that the dovetails were removed for use on other blocks. The dovetails of
wood and stone which are sometimes found with the cartouches of the king
cut upon them were probably inserted after the course had been laid to
serve as a record of the builder which could not be removed by usurpers.
It is certain that a stone dovetail would have little practical value.

In the New Kingdom, the quality of the masonry of walls is, in general,
very poor. Where the walls are solid there is a total lack of internal bond
(Fig. 124). The common form of wall, however, was not solid, but consisted of two parallel walls of masonry with ill-laid blocks or even rubble
between them (Figs. 126 & 127). Even in the parallel faces of masonry,
superfluous half-drums of columns are found, rendering the wall even more
insecure (Figs. 127 & 128). So weak, indeed, are some of the walls in the
most grandiose temples, that they are only saved from falling by the heavy
architraves which rest upon them (Fig. 127). Good hollow masonry is
sometimes met with in late times, the temenos wall in the temple of

Qalabsha in Nubia being an example (Fig. 125); it is about twelve feet thick, the outer skin consisting of good blocks of stone, all laid as stretchers, with dovetails between each pair. The space between the outer and inner skin is filled with rubble. It was an economical method of building an imposing wall, sufficiently strong to resist attacks from the half-civilized people from the south. A thin wall, however, constructed in this manner is very bad practice, particularly if it has to support great weights.

The pylons of the New Kingdom show none of the excellence of fit

Fig. 125. Good hollow masonry of Roman date in the temenos wall of the temple at Qalabsha. It measures 12 feet thick.

within the mass which is seen, for instance, in the packing-blocks of pyramids. Even when the pylon is solid, the laying of the blocks is almost haphazard. The more common form of pylon is, however, far worse from the point of view of solidity, being composed of cells filled in with rough stones. The pylon of Ramesses I (No. II) at Karnak is a striking example of this form of construction (Fig. 130). Here, re-used blocks were so carelessly thrown in that one of the writers was able to crawl in between them nearly to ground level. The settling of the core-blocks has caused the jambs of the great gateway to bulge almost into the arc of a circle. The whole structure is now in a most ruinous condition (Fig. 66). The temple of Soleb, in Nubia, dating to the reign of Amenophis III, had its pylon constructed in almost the same manner as that of Ramesses I. The towers of the pylon were about seventy-eight feet wide at the base, and constructed

Fig. 126. Wall flanking the Great Colonnade of Luxor Temple. The lower part is solid, while the upper part was hollow and filled with rubble. It dates to the reigns of Amenophis III and Tut'ankhamūn

Fig. 127. Remains of a hollow wall in the Ramesseum, showing the poor quality of the masonry. A surplus half-drum of a column has been used at A

Fig. 128. Portion of the north face and the back of the blocks of the south half of a hollow wall of Amenophis III at Luxor Temple, showing surplus half-drums of columns used in the masonry. The south face of the wall, which was never dressed, is shown in Fig. 233

Fig. 129. Western tower of the pylon of King Haremhab at Karnak

of very poor sandstone. The result has been its almost total collapse owing to the pressure exercised on the outer walls by the loose blocks within the cells. Since the pylon was some sixty feet high, this pressure must have been very considerable. The pylon of Haremhab at Karnak is an even worse piece of work (Fig. 129). Its outer skin is only one block thick and it was apparently constructed without any cells. The interior was filled with small

Fig. 130. Sketch-plan of the horizontal section of the north tower of Pylon II at Karnak, built by King Ramesses I. It consists of cells filled in with re-used blocks of King Akhenaten.

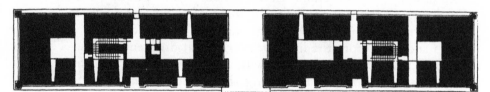

Fig. 131. Sectional plan of the pylon of Edfu Temple, showing the chambers, stairs, and ports to hold the supports of the flagstaves. (From ROCHMENTEIX, *Le Temple d'Edfou*, Pl. I.)

blocks taken from a sun-temple that King Akhenaten had built at Karnak. Though few methods are worse from the point of view of stability than cells filled with a jumbled mass of blocks, a pylon built with stairs and chambers within it can be very strong indeed, since good masonry is ensured almost throughout, and the making of the chamber involves at least a certain amount of internal bonding, while unnecessary weight on the foundations is avoided. Such is the case in the great pylon at Edfu (Fig. 131). Its excellence, however, must be credited to foreign influence rather than to Egyptian building practice.

Having examined the normal details of Egyptian masonry, it may not be out of place to cite a few abnormal examples, though space forbids anything like a complete examination of the strange forms occasionally encountered. At Saqqâra, for instance, the corner of the back wall of what is believed to be the festival temple of King Zoser is laid in the arc of a circle of about thirty feet radius. No other wall of such a form has been found elsewhere in Egypt.

Occasionally, masonry is seen with courses laid on beds concave to the horizon, following the practice in brick (p. 211). Examples can be seen in some of the ancient Nubian houses at Qalabsha (Fig. 132) and Tâfeh. Such forms, however, are not known before Ptolemaic times, and are not

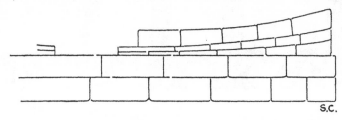

S.C.

Fig. 132. Masonry laid on a concave bed in a house ruin at Qalabsha.
Roman date.

used in temple construction. In the masonry with oblique joints ('Type A', Fig. 105) a variant is met with in some of the queens' pyramids of the IVth dynasty at Gîza, where the masonry is laid on beds which slope down and back from the line of the course at different angles. The joints between the blocks laid on beds with a different slope appear to be at right angles to the vertical plane passing through the front line of the blocks. These pyramids have now been excavated and studied with great care by the Boston-Harvard Expedition (Dr. G. Reisner), and a great deal of valuable information on oblique joints and on the Gîza masonry in general may be expected when the results of this work are published.

X

PYRAMID CONSTRUCTION

THE manner in which pyramids were constructed has long been a disputed point among scholars. This applies not only to the technique of dressing and laying the blocks, but also to the order in which the component parts were built. A pyramid is not just a piled mass of stone, but rather a structure involving many peculiar features not seen in pylons and walls, all of which must carry a meaning if only they are read aright.

The student of pyramid construction is, at the outset, badly handicapped, not only by the lack of published information which is in any way reliable, but by the fact that so few of the pyramids—especially the oldest and most important—have even been properly cleared of the débris which surrounds them. The sides of the Great Pyramid are still largely encumbered with rubbish, and this applies also to the Step Pyramid of Saqqâra and to the pyramid of Meydûm. To add to the difficulties of understanding the development of the pyramid, the builders of two, at Dahshûr, have not yet been definitely identified, though it is quite likely that they may be of very early date.

It must not be imagined that literature on pyramids is non-existent. There are valuable works on the pyramids of Meydûm,[1] of Gîza,[2] and those of the Vth dynasty at Abusîr,[3] but no work of any merit at all has appeared on pyramids in general, nor have the details and peculiarities of their masonry been viewed in a constructive sense.

Now that we have reviewed, as far as possible, the manner in which a stone building was constructed in ancient Egypt, it seems very desirable to try to apply the information gained thereby, with a view to attempting to understand the general principles underlying the construction of a pyramid, though a complete inquiry must necessarily wait for the clearance and competent examination of all the Old Kingdom pyramids.

It is not proposed, in the following pages, to give statistics on the size of these colossal structures (see frontispiece) nor on their internal plans; such information may be obtained in any guide book; neither does it appear necessary to give a *résumé* of the theories of some of the older writers, still

---

[1] PETRIE, *Medum* (1892).
[2] PETRIE, *The Pyramids and Temples of Giza* (1885).

[3] BORCHARDT, *Das Grabdenkmal des Königs*, (a) *Sahurēʿ*, (b) *Newoserrēʿ*, (c) *Neferirkerēʿ*.

less of those of modern cranks,[1] which have already been definitely disproved. It will suffice to review the theories which still have supporters.

Before an attempt is made to determine the manner in which pyramids were constructed, the methods suggested by various authors must be shown either to lead to an impossible situation, or to involve methods with which there is reason to believe that the Egyptians were not acquainted. Two other points must also be borne in mind; one is that the evidence discussed in the foregoing chapters has tended to show that the Egyptians carried out their great building works with the most primitive appliances, and, secondly, that it would be quite unjustified to assume that there is any radical difference between the method of constructing a pyramid and that used for any other structure, except in so far as must result from the pyramid's peculiar shape.

The pyramid seems to be a sun-emblem,[2] and in those of Sneferu at Meydûm and of Khufu at Gîza the proportions are such that, if a circle be imagined whose circumference is equal to the perimeter of the base of the pyramid, the radius of that circle will be its height. This gives an angle of 51° 51' for the angle of the casing (or $\tan^{-1}\frac{14}{11}$).[3] There is no reason to believe that the Egyptians were in any way aware of the incommensurable nature of π.

The development of the Egyptian pyramid can be traced with a certain amount of clearness. The earliest known is the Step Pyramid of Saqqâra, which has definitely been attributed to King Zoser of the IIIrd dynasty. It is not square, but oblong in plan, and seems to show evidence of changes of design. What is clear, however, is that it consists of skins of masonry, one within the other, each skin being apparently faced throughout down to the ground level, the masonry between successive skins being of poorer quality. The angle of the faces of these skins, or 'accretion faces' as they are commonly called, is approximately a rise of 4 on a horizontal distance of 1, this being the angle found on most of the mastaba faces in the Old Kingdom. This pyramid has been described as a compound mastaba, and its construction has been vaguely explained by saying that there was a primitive mastaba which has been successively heightened and enlarged.

The next pyramid (Fig. 133) seems to be that of Meydûm, which was built for King Sneferu of the IVth dynasty, though he is believed to have had another at Dahshûr. Whether the Meydûm pyramid was ever com-

---

[1] These are admirably and amusingly discussed and refuted by BORCHARDT, *Gegen die Zahlenmystik an der grossen Pyramide bei Gise.*
[2] BREASTED, *Development of Religion and Thought in Ancient Egypt,* pp. 11, 15.
[3] Other pyramids show considerable deviations from this angle.

pleted is still a matter of conjecture. Unlike the Step Pyramid, it is square
in plan and of an accuracy of squareness of base and of level not much inferior to that of the Great Pyramid (p. 62). It shows, with certain variations,[1] the same internal structure as that of the Step Pyramid, but it was
covered with an outside facing of very good masonry at an angle of 14 on
a rise of 11, thus being, as far as we yet know, the first true pyramid.

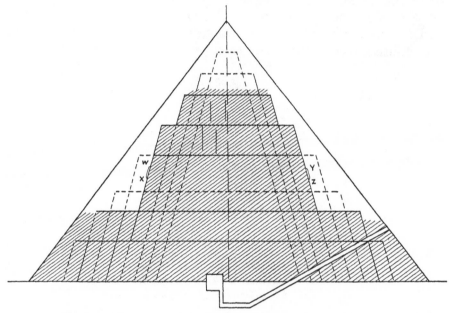

Fig. 133. Internal structure of the Pyramid of Sneferu at Meydûm.
(After Petrie, *Medum*, Pl. II.)

It might well be imagined that, once the form of a pyramid had been
established, the internal skins of masonry would disappear, but they do not.
Though the Gîza pyramids of Khufu, Kha'frē' and Menkewrē' are not
sufficiently destroyed to make their internal construction certain, all the
queens' pyramids of the same date on the Gîza plateau show internal faces,
and so do the pyramids of the Vth dynasty at Abusîr. The tendency
among students is to assume that these 'accretion-faces' played an essential
part in the process of building. For example, in Hölscher, *Das Grab-*

[1] In each 'Step' of Zoser's pyramid there are two
parallel skins of masonry. Further, the 'core-blocks'
behind some of the skins of this pyramid are at right
angles to the sloping face, which is not the case in
that of Meydûm, where they are on level beds.

denkmal des Königs Chephren (frontispiece), a pyramid is shown under construction in the form of stages, brick ramps leading up against the masonry from one stage to another (Fig. 134). This has been reproduced in BREASTED, *Ancient Times*, together with a brief explanatory note on this suggested method of construction.

Before discussing the merits of certain of the modern theories of pyramid construction, two points established in previous chapters must be given the fullest consideration; first, that in Egyptian building, the blocks were handled from the front during the process of laying (Chap. VIII); secondly, that no lifting tackle, other than levers, was used (pp. 43 & 85).

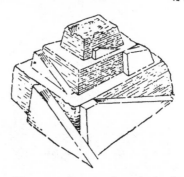

Fig. 134. Suggested method of constructing a pyramid by ramps running from stage to stage. (After HÖLSCHER, *Das Grabdenkmal des Königs Chephren, frontispiece.*)

The theory that pyramids were constructed in stages, with ramps running from stage to stage against the masonry, has, for its main objection, the fact that the blocks could not then be handled from the front while they were being laid. Further, it leaves unsolved the question of the method of putting on the casing.

Another theory, which, strangely enough, still has supporters, is based on a statement of Herodotus, which is of sufficient interest to quote at length. In Book II, Cap. 125, he states:

'This (the Great) pyramid was built thus; in the form of steps. . . . When they had first built it in this manner, they raised the remaining stones by machines made of short pieces of wood: having lifted them from the ground to the first range of steps, when the stone arrived there, it was put on another machine that stood ready on the first range, and from this it was drawn to the second range on another machine; for the machines were equal in number to the ranges of steps (or, they removed the machine, which was only one, and portable, to each range in succession, whenever they wished to raise the stone higher, or I should relate it in both ways, as it is related). The highest parts of it, therefore, were first finished, and afterwards they completed the parts next following; but last of all they finished the parts on the ground and that were lowest.'

The finding of the wooden appliances known as 'rockers' (Fig. 89) in foundation deposits of the New Kingdom, has been seized on by certain scholars to prove that the method outlined by Herodotus was that actually used. Let it be assumed that there were rockers, in unending numbers, strong enough to support blocks up to 10 tons in weight, and that the blocks were rocked up step by step and laid. When the top was reached, the appearance of the monument would be very much like that of the Gîza pyramids to-day. In the putting on of the facing two possibilities present themselves; either the top casing-blocks were also rocked up and the lower courses in some mysterious manner *slipped in below them*, which is a mechanical impossibility, or that the appearance of the casing-blocks, before their faces were dressed, was also that of a series of steps—for by no other means could they also be rocked up. This is directly contradicted by the appearance of the unfinished casing-blocks in the Third Pyramid (Figs. 99 & 100), and also by all other known examples of unfinished masonry.

The third theory is that a primitive mastaba was successively heightened and enlarged. Such a method is a possibility; but successively to enlarge, or widen, the primitive mastaba means the shifting back both of the supply embankments and the foothold embankments all round the courses after each widening, which would be about the most wasteful method of construction, as regards labour, which could be conceived. Observation of the ancient methods always tends to show that the Egyptian was a most efficient organizer of work, though changes in design, not uncommon on the part of the old architects, must sometimes have involved much extra labour.

Since the assumption that the 'accretion faces' were a necessity for the construction of a pyramid seems to lead to a conclusion which is impossible or unlikely, it is justifiable to ask whether such faces may not have been considered by the Egyptians an aid to the stability of the structure, particularly since it is known that the core of all of them, apart from the accretion faces, is of rough masonry, or even whether it was their intense conservatism—or even a religious motive[1]—which made them retain the mastaba angle in the internal masonry of their royal tombs. It is possible

---

[1] The fact that the internal faces of the earlier pyramids were dressed smooth—a perfectly useless proceeding—makes it necessary to take this view into consideration.

that both these factors played a part. Once this is admitted, the problem of
the order of construction becomes very much more simple, for it can then be
assumed that a pyramid was, as a rule, laid out to its full length and width
at the outset, and that the building process consisted merely of heightening
parallel skins of masonry. Thus there would be no moving back of the
main supply- and foothold-embankments at all.

In the pyramids of Zoser and Sneferu, the 'accretion faces' are dressed
smooth, and the masonry is of nearly the same quality as that of the outer
casing; on the other hand, in the queens' pyramids at Gîza and in the pyramids of Abusîr, the 'accretion faces' are not dressed at all (Fig. 135). Here,
at the outset, is a feature in the earliest pyramids which was discontinued—
as far as is yet known—in all the later pyramids. The explanation of the
discontinuance of this practice seems to be that, in order to dress the accretion faces smooth, they had to be kept at a considerably higher level than
the rest of the course, and that, if the dressing of these faces was abolished,
the whole pyramid could be built course by course—or nearly so—
throughout.

For the sake of definition, the masonry of a pyramid may be divided into
four classes; these are (Fig. 135): (1) the core, which lies between successive
'accretion faces', and which is of rough masonry; (2) the internal or 'accretion' faces; (3) the packing-blocks between the outermost accretion faces
and the casing-blocks; and (4) the casing-blocks.

If the internal faces of a pyramid are to be dressed, a considerable number
of courses would need to be exposed at a time, since the dressing of each
individual course to a batter of 4 on 1 would result in bad cumulative
errors. In the Meydûm Pyramid, the error out of the square of one of the
internal faces is as much as two feet.[1] It need not therefore be assumed that
more than some twenty feet of a facing was ever exposed. A confirmation
of this is that, though it was obviously intended to dress all the internal faces
of this pyramid, a series of eight courses of blocks (Fig. 133, *WX, YZ*)
was left undressed in what is now the uppermost face but one. This
can perhaps be explained by assuming that it was soon to be covered over
with the core-blocks, and the omission may well have been due to lack of
inspection or to haste in that period of the construction.

It has been shown that embankments in front of the course are necessary
in masonry involving blocks of any size, and where good laying is required,
and this must have applied to the inner facings of the pyramids as well as
to the casing-blocks, particularly if they were to be fine-dressed. Thus if,

[1] PETRIE, *Medum* (1892), pp. 6 and 7.

for example, ten courses of internal facing had to be constructed above the
level of the main course of the pyramid, there would have to be secondary
embankments all round each face. These would have to be moved back for
each successive facing and so on until they became part of the main embankments round the pyramid. This would be a waste of labour, but not to
be compared with that involved in moving back the main embankments.
If, on the other hand, the internal faces were to be left rough, the construction would be much simplified, since the whole pyramid could then be

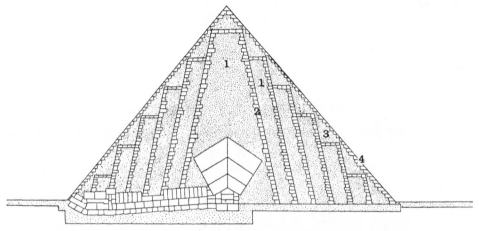

Fig. 135. Section of the pyramid of King Sahurē' of the Vth dynasty at Abusîr.
(After BORCHARDT, *Das Grabdenkmal des Königs Sahurē'*, p. 29.)

constructed almost course by course, the angle of the 'accretion faces' being
maintained accurately by measurement in from the outer casing (p. 125).
It is therefore possible that the Egyptians, perhaps after their work on the
pyramid of Meydûm, dispensed with the useless fine-dressing of the internal faces in order to simplify their task; it is a development in pyramid
masonry which is quite comprehensible.

It may justifiably be asked whether the Great and Second Pyramids of
Gîza possess internal facings. By all analogy they should, since every pyramid of the Old Kingdom before and after them has these facings, as far as
can be gathered from those which have become sufficiently destroyed to
permit an examination of their interior. It has been suggested by certain
scholars[1] that the presence of internal facings is proved by the existence of
the so-called 'girdle-blocks' in the ascending passage leading to the grand

[1] BORCHARDT, *Gegen die Zahlenmystik* . . . , p. 6.

gallery. This passage passes, at intervals of 10 cubits (17 feet 2 inches), right through the middle of single blocks of limestone, and partly through the blocks of masonry between. The passage has been driven through already laid masonry, and it has been held that the 'girdle-blocks' are the faces of internal accretions. Four arguments can be brought against this theory; (*a*) the relative closeness of successive faces, if evenly spaced through the pyramid, would amount to there being no less than fifteen of them, which is at variance with the proportions of the internal facings in all the other known pyramids; (*b*) if the passage were cut at a slope through already laid masonry—was a change of plan, in fact—it would hardly be expected to pass through the middle of a block each time it met one of the internal facings; (*c*) it is against all analogy that the masonry between successive facings should be good, and in this case the masonry is wonderfully fine, though in the places where the true core-blocks can be seen they are, like those of all other pyramids, of rough, ill-laid blocks; (*d*) most of the rising joints between the girdle-blocks and those next to them are very nearly vertical, and some are perfectly so, which would certainly not be the case if the girdle-blocks were the faces of internal 'accretions'. The purpose of this unique and extraordinary method of construction is quite unknown.

If a pyramid were built in the manner suggested, the passages and chambers would be constructed by means of small subsidiary embankments, the blocks forming the sides of a passage, for example, being laid to a convenient height above the general level of the course, and the roofing being put on and the passage extended as the main course became higher. If a passage with the girdle-blocks described above were deliberately sought for, no particular difficulty would be encountered; the Egyptians were no strangers to driving a tunnel through rock and making it straight. The blocks would, however, have to be *arranged* to give the 'girdles' at regular intervals; it is most unlikely that this is the result of chance.

On the order of laying of the core-blocks—the more or less rough masonry which forms the main bulk of a pyramid—information is very scanty. It appears, however, that the principle of cutting down into the rock or into already laid masonry, observed in pavements and the later walls, to receive that of the course above, was freely used.

How the Egyptians dressed the enormous surfaces which form the faces of the pyramids, how they made them meet accurately in a point, and how they avoided twist, are problems rather difficult of solution. The most that can be done is to attempt to determine the principles underlying the process without entering into too many refinements. It must be realized that, to put a batter on the face of a piece of masonry, a plumb-line must be used; further, it must be remembered that in all unfinished masonry in Egypt, which has come down to us, the blocks were laid with their faces rough and fine-dressed afterwards. This dressing could be carried out in two ways, either by covering the whole surface with scaffolding, or by reducing it to the plane of facing-surfaces (p. 62), made when the masonry was being constructed, which would presumably be done, in the case of a pyramid or pylon, when the constructional embankments were being removed. Scaffolding could only have been used for work of moderate size. In the case of the Great Pyramid, each face is some five acres in area, and rises to a height of 160 yards, and to cover such an area with scaffolding would be impracticable mechanically, especially since no traces of 'pot-lugs' are found in the casing at the top of the Second Pyramid at Gîza or in that of the South Pyramid of Dahshûr into which the butts of the scaffold-poles could have engaged, and the whole kept steady. During the removal of the embankments, however, light scaffolding, with the butts of the poles buried in the rubble, may well have been used to cover a height of some thirty feet above them, since it cannot be believed that the dressing of the blocks to the facing-surfaces, if done little more than a man's height at a time, could give the almost perfect flatness observed in the pyramids whose casing has been preserved.

Twist in the pyramid could have been avoided by a very elementary method, namely by establishing points outside the base of the pyramid in line with both diagonals and axes, and projecting these lines up the embankments on to the course under construction by sighting-poles or plumblines. Great accuracy is possible by such methods. The obtaining of the true pyramid angle is however a much more difficult matter. Two possibilities present themselves; one is that the size of the square at the height of the course was calculated, and a square of the calculated dimensions was described on the already established diagonals and axes; the other is that it was found by plumbing and measurement. In the records left to us by the ancient masons, dimensions and levels are indicated by a triangle (generally in red ochre) with either the base projected out or a second line drawn parallel to the base,[1] accompanied by indications of the number of cubits, palms, and digits (p. 63). In such marks the base of the triangle or the parallel line usually shows whence the measurement was to be taken and the apex of the triangle the direction of measurement. These levels show no great pretensions to accuracy. Since there is no evidence, from any of the partly

---

[1] BORCHARDT, *Das Grabdenkmal des Königs Newoserrē*, p. 154.

destroyed pyramids, that vertical pits were left in the masonry from which the height of the work could be accurately obtained at any course, and since the use of the rough level-records found on the blocks would involve very large cumulative errors, it is fairly safe to assert that the size of the square at any particular course was not calculated. The alternative solution is that a plumb-line was used in the embankments outside the course and the necessary proportion of 11 to 14 measured in from it. From these measurements, the facing-surfaces could be established to which the whole of the casing would afterwards be dressed when the embankments were being removed. This seems to be the only solution which meets the observed facts, and it is worth examining more closely. In the case of the Great Pyramid, it appears that the error of the square at what is now the top cannot have exceeded a foot. If the angle of the casing for each of the 200 courses had been separately obtained by plumbing, the observed accuracy could never have been obtained. From the writer's experiments, it appears that plumbing cannot be carried to an accuracy of much more than $\frac{1}{5}$ inch with the Egyptian plumb-bobs which have been preserved, even if the plumb-line is sheltered from the wind and all the obvious precautions taken. With 200 repetitions, the cumulative error could well be three times as great as that observed. If, on the other hand, the plumbing-point on any course could be kept open for some seventeen feet, only about thirty shifts of the plumbing-point would be necessary during the construction of the pyramid, and the observed accuracy could well be attained.

Since it appears likely that the plumbing was done from a point outside the face of the pyramid and that this plumbing was not done consecutively at each separate course, it must necessarily follow that there were pits or trenches in the embankments. What the exact form of these pits or trenches was we have no idea, neither can we determine how many of them were used for each face.

The foothold-embankments round the course under construction cannot have had a breadth, at the top, of much less than forty feet, if parties of men had to manipulate large levers for the laying of the casing-blocks, and it is not without interest to determine how far, by means of pits or trenches in the embankments, the plumbing could be carried before a new plumbing position would have to be used. Since the angle of the casing of a pyramid is a rise of 14 on a horizontal distance of 11, if the pit were commenced close to the face, it would be fifty-one feet deep by the time it reached the edge of the embankment higher up. Thus only ten shifts would be necessary in the construction of the Great Pyramid. Although these figures are merely tentative, they serve to show that the observed accuracy could be obtained by simple methods. A practical application may make matters more clear.

Let it be imagined that a pyramid is under construction and that the base has been set out to the necessary accuracy (p. 65), and, after laying the blocks of the first course (p. 109), the line of the base has been exposed (where the edge of the casing will touch the pavement) at, say, two points on each side and some forty feet in from the corners, the top of the course having been truly levelled. Having measured in from a plumb-line over the exposed points on the base, on the first course, two points can be ob-

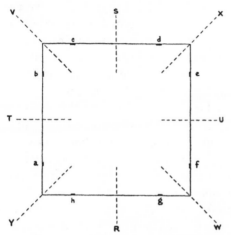

Fig. 136. Diagram illustrating suggested method of dressing the faces of a pyramid.

tained on each side (*a b, c d, e f, g h*, Fig. 136) on the plane to which the whole of the surface of the face will afterwards be reduced, by measuring in $\frac{11}{14}$ of the length of the plumb-line. From these points, which become the facing-surfaces of the course (p. 62), a square can be established which can be checked by measuring its sides (and sometimes its diagonals) [1] and by ascertaining if it lies on the diagonals of the pyramid which have been projected up on to the course (*V W, X Y*), as suggested on p. 125. The accuracy of the square can be further checked by ascertaining, by measurement, if the axes of the pyramid base, when also projected up on to the course (*R S, T U*), pass through the middle of each side, and thus avoid the possibility of the would-be square becoming a rhombus. With such pre-

[1] In the Great Pyramid, for a considerable number of courses, the diagonals could not be measured, since a mass of rock rises up into the body of the pyramid to an unknown height.

cautions, a square of very great accuracy is possible. The process would be repeated for the higher courses from the same plumbing-point on the pavement, until the top of the plumb-line had reached the extreme edge of the foothold-embankment; it would then be shifted in to the exposed facing-surface of the course last laid.

In the case of the pyramids, there is no evidence on the precise form taken by the facing-surfaces before they were covered up by the constructional embankments for the next courses. In later times, two forms are known (Chapter XVIII).

In the Third Pyramid at Gîza the sixteen lowest courses are cased with granite blocks (Figs. 99 & 100), which were never dressed, though there is evidence to show that the dressing of the limestone casing—now broken up—which covered the upper part of the pyramid was completed. It is likely that the granite courses were left unfinished owing to the enormous labour necessary to dress the hard rocks (p. 27). The granite casing has no visible facing-surfaces on it; perhaps the intention had been to dress the granite courses, at leisure, to the plane of the finished limestone casing above. Some base must however have been set out at ground-level, and since the granite blocks show no trace of facing-surfaces on to which plumbing could have been done, we can infer that the Egyptians were capable of plumbing, in one operation, for a height of at least fifty-four feet up a pyramid, which would mean that the top of the plumbing-pit, or trench, could be forty-two feet out from the face of the masonry. It will be seen that these figures agree with the hypothetical calculations on p. 126.

Most pyramids have individual peculiarities which are as yet difficult to explain. For instance, in the Great Pyramid, as possibly in certain others, a large depression in the packing-blocks runs down the middle of each face, implying a line of extra thick facing there. Though there is no special difficulty in arranging the blocks of a course in such a manner that they increase in size at the middle, there is no satisfactory explanation of the feature any more than there is of the 'girdle-blocks' already discussed. To assume that vertical lines of blocks were laid from the top of the pyramid to the bottom, to serve in some manner as a guide, is incompatible with the deduction that large embankments were used, without which many of the peculiarities of ancient Egyptian stonework are inexplicable. Another reason which can be advanced against such a strange order of laying having taken place is that the height of the casing-blocks of the Great Pyramid shows a periodical decrease for a certain number of courses up the face, which must surely mean that the available stone (or perhaps that from one quarry) was worked out before a new supply was drawn upon. The assumption that such vertical lines of blocks were laid would mean that the amount of periodic decrease in the height of the courses was exactly foreseen. Other considerations, such as the levelling of the tops of the courses after laying, are also against such a supposition being true.

In conclusion, the reader is again reminded (see the Preface) that the foregoing notes on pyramid construction are not to be regarded as a complete and final exposition of the many problems hitherto unexplained, but rather as preliminary deductions which seem to follow from the information at present available, and which may have to be considerably modified in the light of future research.

XI

PAVEMENTS AND COLUMN-BASES

IN constructing a pavement, the Egyptians did not generally make use of rectangular slabs; the whole system is a patchwork (Fig. 137), the blocks being laid to their full depth in recesses cut into the foundations to

Fig. 137. Part of the pavement in the valley temple of the Second Pyramid ('The Temple of the Sphinx') at Gîza; IVth dynasty. (After HÖLSCHER, *Das Grabdenkmal des Königs Chephren.*)

receive them, so that their rough upper surfaces should all lie at about the same height. The top of the pavement-slabs was smoothed and levelled after all the blocks had been laid. That the dressing of the top of the slabs was, like those of the blocks in a course, done last, can be easily deduced from a careful study of Egyptian pavements, even including those of the latest periods. For instance, in the Temple of Edfu, where, in the Hypo-style Hall, the level of the pavement between the middle ranges of pillars is about five inches below the remainder, blocks may be seen at the step dressed partly to one level and partly to the other. In some buildings, the pavement seems to have been laid last of all; at Philae, for example, the walls have in some cases been cut to waste to receive the paving-slabs (Fig. 138).

In the case of the monolithic pillars which required no base, as, for example, in the temple of the Second Pyramid, the pavement, after being laid, was cut away so that the pillars could be passed down through it to rest on the foundation. In other structures, such as the 'Osireion' of Seti I at Abydos, the huge square granite pillars rest on the pavement.

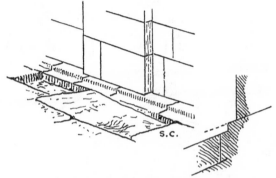

Fig. 138. Part of a wall in the temple of Isis at Philae, showing cutting to waste for the laying of a pavement and for the formation of a pilaster.

The Egyptians never gave very great attention to the bases of their columns, either from a decorative or from a structural standpoint; they are either entirely absent or are of the simplest forms, often rising very little above the floor-level. As examples of the curious methods of construction evolved by the Egyptians they are, however, of considerable interest, and show several varieties of technique, few of which can be definitely assigned to one period. They fall roughly into two classes. In the first, the bases stand on already laid and flattened paving-blocks. In the second, the base-blocks rest on the same foundation as the pavement, and may either be square or circular in plan or of the same irregular form as the other paving-slabs.

The first class, of which good examples can be seen in the temple of Luxor, does not call for much description. Structurally they are bad, as the pavement-slabs are liable to be split under the weight unless they are of very great thickness.

The second class shows many varieties. The granite column-bases which supported the great monolithic columns in the north pyramid temple at

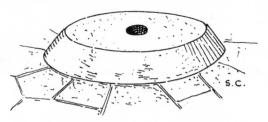

Fig. 139. Granite base for monolithic column in the northern pyramid temple at Abusîr. The basalt paving-blocks butt against it. (Vth dynasty.)

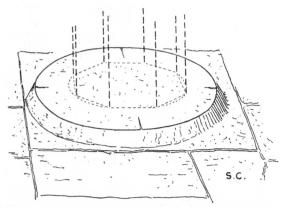

Fig. 140. Base for one of the octagonal columns in the XIth dynasty temple at El-Deir el-Bahari, showing marks left by the masons as a guide for the setting of the column. The columns are 22 inches diameter.

Abusîr of the Vth dynasty rest on the foundations of the temple, and the basalt pavement butts up against them (Fig. 139). A more common method of construction was to cut the base from a block which was nothing more than a paving-slab of greater height than the rest, and which was laid with those of the rest of the pavement. In the XIth dynasty temple of El-Deir el-Bahari, the base of the octagonal columns forms part of a more or less square block, which only rises about 1½ inches above floor-level (Fig.

140). In the adjoining temple of the XVIIIth dynasty the masonry is of
a poorer quality, and columns, weakened by patches, can be seen, whose

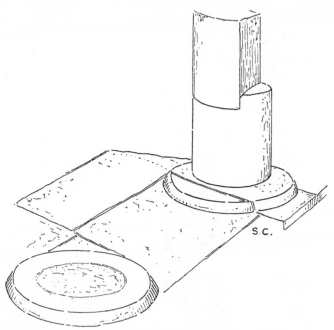

Fig. 141. Column-bases in the XVIIIth-dynasty temple at El-
Deir el-Bahari. These are only paving-slabs which have been cut
down less than those of the remainder of the pavement.

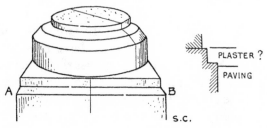

Fig. 142. Foot-stones below the base of a column at
Edfu.

bases sometimes form part of two blocks, showing that they were merely
ordinary paving-blocks cut down less than the remainder (Fig. 141). In
the Temple of Edfu, the column-bases sometimes rest on squared blocks

which lie on the foundation (Fig. 142). The pavement does not come up
to the level of the square base-blocks, a ledge three inches deep (*AB*) being
left, which shows either that there was another layer of blocks over those of
the present pavement or, and this is more likely, that the difference was
made up with a thick layer of plaster.

An interesting column-base is recorded from Abydos in the area known
as the 'Kôm El-Sultân' (Fig. 143). It dates to the reign of Ramesses II,
and is composed of one block of sandstone 2 feet 6 inches high. The upper
part is rounded and the lower part square, the sides of the latter measuring

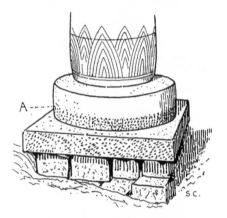

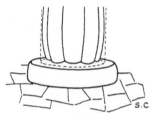

Fig. 143. Sandstone base of a limestone
column at Abydos, the height of the ancient pavement being shown by that of
the rough tooling of the block. (*A*.)

Fig. 144. Circle scored on the
base of a column, exposed after
the diameter of the lower part
of the column had been reduced. Karnak.

7 feet 3 inches. The whole base is roughly tooled to within a foot of the
top, showing that the pavement level must have been at that height, thus
resting on part of the base and butting against the remainder. The foundations of this base were of insignificant blocks of limestone, set together with
poor mortar and stone chips. The shaft of the column was also of limestone.

Guide-marks, scored into the stone with a pointed tool, are frequent on
base-blocks, and they are sometimes seen on the paving in cases where the
base rests on the pavement. They usually take the form of two lines at right
angles, or four nicks, and were intended to mark the axis of the column
or base. Occasionally a circle is seen on the base-blocks, which may either
be a line down to which the column was to be dressed after erection or else
the marks of the tools used during the dressing. In the Hypostyle Hall at

Karnak the foot-stone of the columns is scored with a circle (Fig. 144) whose diameter is the greatest diameter of the column. Though it may have been a guide-line for the masons, it is more likely that the bottom of the column was originally of its present greatest diameter and that the line was left by the tools in the dressing when it was sculptured and its size reduced. The use of these guide-lines is discussed in Chapter XII.

XII

COLUMNS

COLUMNS, in Egypt, may be roughly divided into two classes, both of which can be traced back to very remote times. The first type imitates, in stone, such vegetable growths as bundles of reeds tied together and stiffened with mud, palm-trunks, &c.; the second type is the square-sectioned pillar with its later derivations, which may possibly have had its origin—if an origin indeed be sought for such a simple form—in the supports left to sustain the roof in quarries and rock-tombs.

The great difficulty with which the Egyptians had to contend was to roof a chamber of any considerable size. In the primitive huts, on which so much of their architectural style is based, a few posts will keep up quite a large area of reed and palm-frond roofing, but to cover a space with stone requires a great number of supports. It may justly be said that in the Great Hypostyle Hall at Karnak one cannot see the hall for the columns.

In the IVth dynasty the greatest space which could be spanned with limestone blocks was just over nine feet; larger chambers had either to be covered with granite or by some form of pent or corbelled roof (Figs. 218 & 219). When the Silsila sandstone quarries were brought into general use it was found that this stone could be used for very much longer architraves. The Egyptians were, no doubt, sensible of the need for reducing the number of columns as far as possible, and we find them using sandstone to span a distance which, in the central aisle of the Hypostyle Hall at Karnak, amounts to nearly thirty feet. Although an excellent material for general building purposes, sandstone is not capable, when used over such distances, of bearing roofing-slabs for long in addition to its own weight, and it can hardly be doubted that the architraves and roof-slabs at Karnak began to give way within three centuries of the completion of the building.

The square-sectioned column is known from the IVth dynasty until the end of the New Kingdom, both in its monolithic and built-up forms. During the Middle and the early part of the New Kingdom, it underwent, at times, a development into the polygonal column, which is both simple and pleasing. The first step in this development must have been the cutting off of the angles of the square column (Fig. 145), which produced the octagonal form (Fig. 146) which is found in the XIth dynasty temple of El-

Deir el-Bahari and elsewhere. A further cutting of the angles resulted in a sixteen-sided column (Fig. 147), which occurs in many temples and tombs both in the Middle and the New Kingdoms.

Champollion, impressed by the polygonal column, and remembering that the Greek Doric column (Fig. 148) is fluted, christened the Egyptian

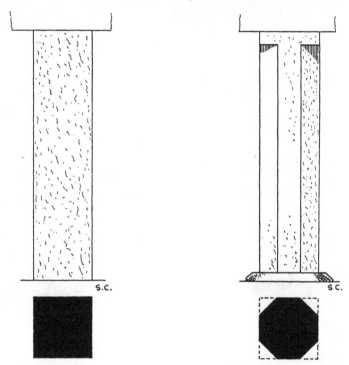

Fig. 145. Square-sectioned column.   *Fig.* 146. Octagonal column. XIth dynasty; El-Deir el-Bahari.

column of the polygonal type 'Proto-Doric'; but Egyptian columns are not all fluted, in fact they more often have flat facets. In other respects the term 'Proto-Doric' is a misnomer; in the Greek form there is a very distinct taper upwards, and at the top it is furnished with a large overhanging capital made from a block of stone separate from the shaft of the column. In the Egyptian column the 'abacus' does not project all round; the four cardinal faces of the column are almost on the same plane as the corresponding sides of the block above it.[1]

[1] In some of the most serious works on Egyptian architecture the draughtsman commits the error of representing the block of stone above the column projecting all round the capital.

A variation in the treatment of the polygonal column was to leave slight ridges between successive facets (Fig. 149). In examples in the temple of

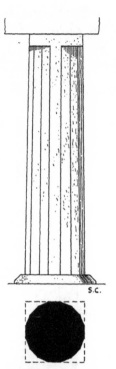 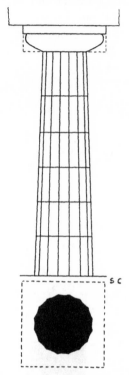

Fig. 147. Sixteen-sided column, XIIth–
XVIIIth dynasty.

Fig. 148. Greek Doric column.

Fig. 149. Plan of part of a polygonal column, showing ridges between
the facets.

Khonsu at Karnak, when examined by one of the writers some years ago, there were traces of the plaster on the facets, not filling up the gap between successive ridges, but lying flat on the stone. Such ridges, though known in other temples, are comparatively rare.

Fig. 150. Polygonal column surmounted by a mask of the goddess Hathor in the small temple of Amenophis III at El-Kâb

One of the most effective examples of the ornamentation of the poly-gonal column is seen in the little temple of Amenophis III at El-Kâb (Fig. 150). The four columns which support the roof are sixteen-sided, and, facing the central nave, a vertical band occupies the space of two facets.

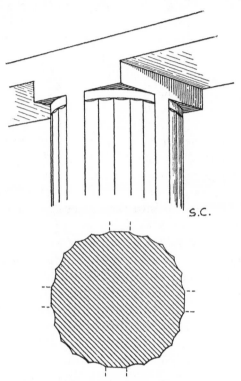

S.C.

Fig. 151. Fluted column in the rock-hewn temple
of Amenophis III at Beit el-Wâli in Nubia.

At the top there is a mask of the goddess Hathor, with the customary little shrine above it.

Polygonal columns enjoyed but a short life in comparison with those derived from vegetable sources. It seems that they did not please the vulgar minds of Ramesses II and his followers. At any rate, after his time they are no longer found.

Fluted columns are no very great rarity in Egypt. The principle of fluting was known, and freely employed, by King Zoser of the IIIrd dynasty (Figs. 1 & 2), though no free-standing fluted columns have been yet found in his work. In the Old Kingdom they seem not to have been used after Zoser, as far as we yet know, but they reappear again in the Middle Kingdom at Hawâra[1] and 'Kahûn'[2] and in some of the rock tombs at Beni Hasan. In the New Kingdom the only temple in Egypt in which a fluted column has been found to our knowledge is that of Mût at Karnak, where, among the ruins, there are a base and a top-drum below the capital

of a truly fluted column. Four bands are carried up the cardinal faces, and between each pair of bands are seven slightly hollowed flutings. The column is about thirty-eight inches in diameter. At the temple of Sedenga, in Nubia, a column is still standing with thirty flutings and two vertical strips bearing inscriptions of Amenophis III. Another at Beit el-Wâli, which, unlike that of Ameno-phis III, is cut out of the rock (Fig. 151), has the unusual feature of a narrow band below which the flutings begin.

Fig. 152. Section of a papyrus stalk.

Down the four cardinal faces are vertical strips, not fluted but flat, which die at the top into the 'abacus'. The column has a distinct taper, but no bulging or 'entasis'.

Columns imitating vegetable growths are common from very early times. They are based mostly on the papyrus-plant, the lotus or lily and the palm.

One of the most pleasing forms of column based on the papyrus is that imitating a bundle of these plants with their heads left spread out. The shaft is circular in plan, and is slightly reduced where it stands on its flat, circular base, this part of the shaft being carved and painted to represent a group of large leaves pointed at the top. The shaft diminishes as it ascends, and is surmounted by a wide capital which springs immediately above several horizontal lines round the shaft, relics of the bands which tied to-gether the reeds. Though the column thus shows that its prototype was many reeds tied together, there are indications that it was also meant to recall a single plant. For example, up the shafts of some examples, ridges may be observed at every third of its circumference (Figs. 152 & 153); this is a feature of the papyrus stalk. The peculiar section of Zoser's pilaster (Fig. 7), which is crowned by a capital which strongly suggests the papyrus-head, must have had a similar origin. This type of column is unknown, as yet, in the Old and Middle Kingdoms, and does not appear in the New Kingdom until the XIXth dynasty, when it is used for the two middle ranges of columns in the great halls of that period. One of the most

[1] Petrie, *The Labyrinth, Gerzeh and Mazghuneh*,   [2] Petrie, *Illahun, Kahun and Gurob*, Pl. VI. Pl. XXIX.

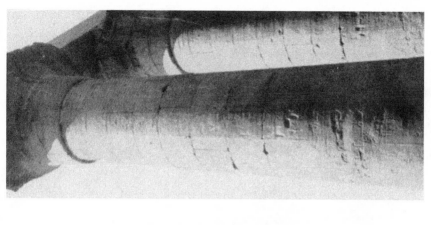

Fig. 154. Ridges on the columns of Tut-'ankhamūn in the Great Colonnade at Luxor Temple. There are three ridges on each column intended to imitate those on a papyrus stalk

Fig. 153. Grand Colonnade of Luxor Temple, commenced by Amenophis III and completed by Tut'ankhamūn

impressive examples is the Great Colonnade at Luxor Temple, which was finished by King Tut'ankhamūn (Fig. 154). Though slightly smaller than the central aisle of the Great Hypostyle Hall at Karnak, and not flanked by rows of lesser columns, it is undoubtedly more attractive. The capitals have a majestically curved outline, and extend far beyond the diameter of the column. The ten columns which flank the central nave of the Great Hypostyle Hall at Karnak have capitals 18 feet across the lip, and are built up of a few large blocks (Fig. 155). One block forms half the top course of the capital, the dimensions of which, in the rough, must have been 20 feet by

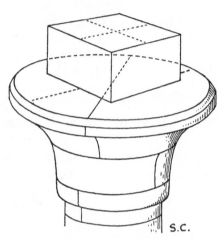

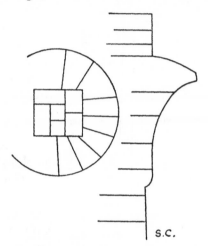

Fig. 155. Capital of a column in the central aisle of the Great Hall at Karnak, showing mason's guide-lines to assist in placing the abacus and the architraves. This capital measures nearly 20 feet across the top.

Fig. 156. Part of plan and sectional elevation of the papyrus capital of King Taharqa, showing how it was built of small stones. (For a photograph of this column see Fig. 69.)

10 feet by 2 feet 6 inches deep. The capital is so large that it was found necessary to construct it of three deep courses of stone. That the Egyptians cut out the form of the column and capital after the stones were in position is shown by the fact that the lower horizontal joint is not at the point of the springing of the capital from the shaft, but below it. The quantity of stone cut to waste in making these capitals must have been very great.

The columns of the Hypostyle Hall at Karnak and those of Luxor Temple are examples of the most solid form of construction, and were intended to carry very great weights. In contrast to these, the beautiful column of King Taharqa, the last survivor of ten which stood before the

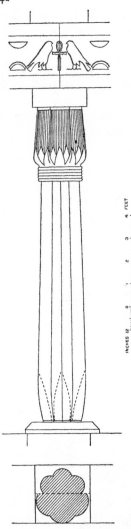

Second Pylon at Karnak (Fig. 69), deserves
attention. These papyriform columns were as tall
as those in the central nave of the Great Hall
of Karnak, but were never intended to carry
any heavy load, being constructed of comparatively small blocks of stone (Fig. 156).[1] Not
only has each drum of the column many more
stones than those already described, but there
are many more joints vertically and courses horizontally. Such numerous joints are comparatively
harmless when no huge architraves or stone slabs
were to rest on the column, but had these columns
been loaded with such weights as were carried by
those of the Hypostyle Hall at Karnak they would
inevitably have collapsed. The capital of Taharqa does not spread to the same extent as those
of Karnak, and the rectangular block or abacus
at the top of the capital is made of quite small
blocks. This range of columns was not roofed at
any time. Fine as they are, their construction is
bad. When newly built, with all the joints filled
with mortar and the whole surface painted and
sculptured, their weakness would not be apparent,
but inundation and perhaps earthquakes have
played fearful havoc with this building.

The second type of papyrus column[2] imitates
a number of bunches of the plant bound together,
the heads being tied in instead of being allowed
to spread. They date from the Vth dynasty;
fine monolithic examples in granite were found
in the pyramid-temple of King Sahurē‘ at Abusîr
(Fig. 157), standing over twelve feet in height.
After the Old Kingdom, the space between successive bundles is filled by small bundles of what
seem to be reeds, cut off short at top and bottom.
Like the other form of papyrus column, the shaft

Fig. 157. Monolithic granite
column of the Vth dynasty,
from the pyramid-temple of
King Sahurē‘ at Abusîr. (After
BORCHARDT, *Das Grabdenk-
mal des Königs Sahurē‘*, Pl. XI.)

---

[1] See also CHEVRIER, *Annales du Service* xxvii, p. 141.
[2] The development of this and the previous column
is well illustrated in JÉQUIER, *Les Éléments de l'Archi-*

tecture Égyptienne, pp. 201–30. Here, the papyrus
column with the spreading capital is termed 'cam-
paniforme' while the second type is 'papyriforme'.

Fig. 158. Mortuary temple of Ramesses II (The Ramesseum) at Thebes, from the north-east. (Photograph by Gaddis and Seif, Luxor)

Fig. 159. Lotus column of limestone, from the Vth dynasty pyramid-temple of King Sahurē' at Abusîr. (Photograph by the Antiquities Department)

is reduced in size near the base and tapers up towards the top. The edge of the bundles which go to make the shaft are all ridged in the better examples. This form of papyrus column becomes the chief type for the temples of the XIXth dynasty,[1] when it degenerates into gouty monstrosities which reach perhaps their most unpleasing shapes under Ramesses II (Fig. 158). In these the modelling of the shaft disappears and the column is covered with decorations and inscriptions in the most unsuitable way imaginable.

The Lotus column (Fig. 159) is contemporary with the second form of papyrus column, both being found in the temples of the Vth dynasty at Abusîr. It obviously represents several lotus buds with their stalks bound together in the same manner as the bundles in the papyrus column; in fact, the two types are very closely allied. In good examples the chief differences are that the lotus column is not reduced in size near the base, and that the bundles are round and not ridged. The papyrus heads, bound together at the top, are in this column replaced by lotus buds, with smaller plants introduced between successive buds under the lines below the capital, which represent the bands with which they were tied. In this feature the papyrus columns of the second form seem to have copied the lotiform column, since, from the tomb scenes, it appears that it was a very common custom to bind lotus buds at the top of the pillars supporting the roofs of houses and shrines. On the other hand, in a more important aspect, it is clear that the lotus column was the successor of the papyrus column, since by no manner of means can the weak stems of a lotus plant be made into a support, however

[1] M. Jéquier calls this type 'monostyle'.

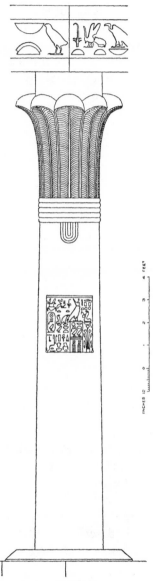

Fig. 160. Monolithic granite palm column of the Vth dynasty from the pyramid temple of King Sahurē' at Abusîr. (After BOR-CHARDT, *Das Grabdenkmal des Königs Sahurē'*, Pl. IX.)

small. The lotus columns are less frequent than the papyrus, judging from the number of examples known to us. They are seen at Beni Hasan in the Middle Kingdom,[1] and are fairly frequent in Saïte and Ptolemaic times.

The column imitating the palm is also found in the Vth dynasty temple of Sahurē' at Abusîr (Fig. 160). Though known in the Middle Kingdom, only two examples are found in the New Kingdom, namely, at the temple of Amenophis III at Soleb and that of Seti I at Sesebi, both sites being in

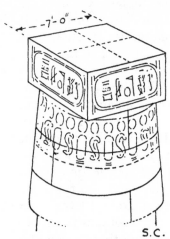

Fig. 161. Mason's guide-lines on the abacus of a column in the Great Hall at Karnak; XIXth dynasty.

Nubia. In Ptolemaic times they become frequent, and it is only in this period that the shafts are made to represent the surfaces of a palm trunk.

Other types of column are occasionally found in Egyptian temples, such as a cylindrical column surmounted by a square abacus[2] and a curious type representing the traditional form of the tent-pole (Fig. 174). A full description of these and the beautiful composite capitals[3] used just before and during the Ptolemaic era is, however, outside the scope of this volume, since their interest lies in their style rather than in the details of their construction.

---

[1] NEWBERRY, *Beni Hasan,* ii, Pl. X.
[2] JÉQUIER, *Les Éléments de l'Architecture Égyptienne,* Fig. 101, from the temple of Sahurē' at Abusîr (Vth dynasty) and Fig. 102 from the temple of

Seti I at Abydos.
[3] A very good selection of these composite columns can be seen in JÉQUIER, *op. cit.*

Fig. 162. Unfinished column, architrave and cornice in the court of
Sheshonq I, at Karnak

In order to assist in the placing of the architraves, lines were drawn, or a series of dots made with a metal point, not only on the abacus, but on the top of the capital, to ensure that the abacus should be properly dressed. Such guide-lines can be seen both in the column of the central nave of the Great Hypostyle Hall at Karnak (Fig. 155) and on the smaller columns that flank them (Fig. 161).

Columns, like walls and pavements, were always dressed after the blocks forming them had been laid. This is proved not only by the frequent incorporation of the base or the capital into the shaft, but also by the examples of columns which have either wholly or partly been left unfinished. Unfinished columns are rare; the best example is seen in the south-east corner of the XXVth dynasty court on the east of Pylon I at Karnak (Fig. 162). This had to wait for the completion of the pylon, since it was covered by the constructional slope by means of which the pylon was being built. For some reason of which we are ignorant this great pylon was never finished, and since the constructional embankments have largely disappeared in the course of the ages we can now study the corner of the court and see, not only an unfinished column and architrave, but the repairs which were under way in the little temple of Seti II beside the northern tower (Fig. 239). In the column, between the bedding joints of the blocks which were to form the capital, the stone has been dressed in rings which were but little larger than the diameter of the corresponding part of the finished column next to it.

The east wall of one of the chambers (F of Baedeker's plan) in the temple of Seti I at Abydos had originally been intended as a row of columns. The blocked-out form of these is still visible (Fig. 163), since the wall, which was made to incorporate with them, has now been broken away. The columns are of sandstone, and were shaped by a round, blunt-ended tool before being finished with a chisel and scraped with a hand-tool like the remainder of the work.

In the drums of columns whose bedding surfaces have become visible, it is comparatively rare to see any guide-lines, though they are by no means unknown. Some of the drums of the columns of the side-aisle in the Great Hall at Karnak have their top surfaces intersected by two lines, at right angles, marked on the stone with a pointed tool, these lines indicating the centre of the drum. It is very probable that the upper surfaces of all blocks for column-drums not only were marked with lines or nicks from which the centre could be found, but also had circles cut on them showing the line down to which the stone was afterwards to be dressed. Unless the axial lines

were continued across the surface of the drum, as at Karnak, all guide-lines
would be absent on the drum of a finished column.

Since the drums were laid with the outer surfaces only roughly shaped—
if at all—and since, like all other Egyptian masonry, they were probably
laid with the top surface also rough, it follows that the only fine-dressing
done on a block for a column-drum would be to flatten one surface as a
bed.[1] When the blocks were laid the top would be levelled and the axis
of the column established. This would almost certainly be done by sighting

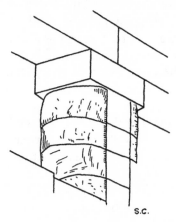

S.C.

Fig. 163. Column in the temple of Seti I at Abydos.
Owing to a change of plan, the line of columns was
converted into a wall. The destruction of the wall
shows the state in which the column-drums were laid.

or running a line down the rank and file of all the columns in the hall and
marking where these lines touched the outer edges of the surface of the
drums. These could be joined, as at Karnak, thus giving the axis, or centre
of the column; or else short lines, or even nicks, would be left, from which,
by means of threads, the centre could be found and the circle of the
requisite size drawn.[2]

The manner in which a monolithic column would be shaped, such, for
example, as one of the granite papyrus columns at Abusîr, is unknown
from any ancient indications, but our knowledge of Egyptian crafts enables
us to limit the possible methods by which they may have done it. Such

---

[1] If the drum were of two blocks, the rising joints
between them would, of course, also have to be
fine-dressed.
[2] We have no information on the Egyptian method

of drawing a circle; most likely the point of the
marking tool was passed through a loop in the end
of a piece of string held at the centre.

a column would arrive from the quarry as a long block, perhaps of more or less rectangular form, and would certainly be erected before being shaped. A base would, however, first have to be dressed on which it could stand, and, further, it would have to be dressed so that a vertical column of the required form could be shaped from it without any further movement or adjustment. There seems but one manner by which this process could be carried out by

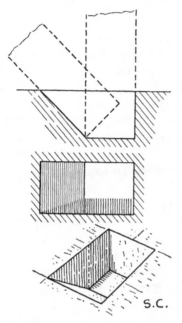

Fig. 164. Hole in the granite pavement of the temple of the Second Pyramid at Gîza. One side of the hole was cut to a slope to facilitate the insertion of the butt of the square pillar.

simple methods. Points on the two extremities of the block would first be chosen to define the axis of the finished column. Careful consideration and measurement would be necessary in choosing these points so that there should be enough stone round the axis to furnish a column of the required shape. The next problem would be to dress the base; that is, to make a plane to which the axial line should be at right angles or 'normal'. A simple method, which may well have been used, would be to lever up the block until the two axial points chosen were exactly level (p. 62) and, at the base-end, to make a truly vertical facing-surface from the top to the bottom of the block. The next step would be to turn the column through a right-angle, re-level the two axial points, and make another vertical facing-surface passing through a point about the middle of the one previously made. The base would then be dressed to a true plane on the two facing-surfaces (p. 105). The block, when erected on this surface, would then stand with the chosen axis vertical.

The method by which a monolithic column was erected cannot be defined too closely, though several possible methods can be suggested. In the temple of the Second Pyramid the square columns have been slipped in through holes in the pavement on to the rock foundation. One side of the hole was cut sloping (Fig. 164), down which the butt of the column, which must have been dressed, seems to have been slipped. Whether the column was got into an upright position by means of levers, or was erected by means of a constructional embankment, as may have been the case with obelisks,[1] there is no means of knowing. It seems likely that levers were used in all but extreme cases. As an example of the last, the gigantic square-sectioned columns of the so-called 'Osireion' of Seti I at Abydos may be cited. The building is below desert level, and is walled by huge blocks of quartzite, and the pillars, which supported a granite roof, were of square section, measuring 13 feet in height with a side of 7 feet 6 inches. It is likely that they may have been let down into the hall by filling the whole of it with sand and removing the sand gradually from under the columns. In this case the pillars rest on the pavement and not in the manner of those of the Second Pyramid temple at Gîza.

The dressing of a monolithic column may have been carried out by marking a circle on the floor or the base or upon a prepared platform and measuring in from the plumb-lines in accordance, perhaps, with a scale drawing of the column. By this means any number of facing-surfaces could be obtained at different heights, though we have no information at all either on their number or form.[2] The forming of pillars in a rock-hewn temple or tomb must have been an even more difficult proceeding than that of a monolithic column, and here again we are very short of any precise indications which might enable us to determine the most likely method adopted. In certain of the XVIIIth dynasty rock-tombs at El-'Amârna, which have been left unfinished, it seems that the floor was the last to be dressed, and partly excavated tombs can be seen where the floor was still undefined, but where not only the roof, but the capitals and part of the shafts of the columns had been dressed. Though several methods could be suggested which would

---

[1] The method by which obelisks may have been erected is fully described in ENGELBACH, *The Problem of the Obelisks*.

[2] ENGELBACH, *Annales du Service* xxviii, pp. 144–152.

explain all the observed facts, it would be unwise to formulate them before a very detailed examination had been made of the accuracy of the rock-hewn columns of El-'Amârna and elsewhere.

The only determination of the accuracy of an Egyptian column of circular section which has yet been made is one on the red granite monolithic columns of the Vth dynasty from Abusîr (Fig. 160). It was found that on a portion 2 m. 60 cm. long, where the mean diameter tapers from 91·2 cm. down to 79·8 cm., the error from the mean diameter was never greater than 8 milli-metres.

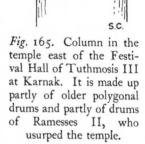

The taper, however, shows a lesser degree of accuracy, which probably means that this column, like the unfinished example at Karnak, was dressed by first forming rings of the desired diameter at regular intervals up the shaft, and then smoothing the parts between them, and not by means of a series of vertical facing-surfaces. It is likely that some form of templet was used, though there is no direct evidence that this was the case.

The Egyptians, particularly in the XIXth dynasty, very frequently made use of stone from an earlier building when constructing a new temple. More rarely, column-drums were re-employed, and new columns constructed from them. East of the Festival Hall of Tuthmosis III at Karnak there is a temple appropriated by Ramesses II. In this building it can be seen that some of the columns are entirely composed of re-used polygonal drums, others are

Fig. 165. Column in the temple east of the Festival Hall of Tuthmosis III at Karnak. It is made up partly of older polygonal drums and partly of drums of Ramesses II, who usurped the temple.

partly Ramesses II's and partly polygonal drums (Fig. 165). They were well covered with a thick coat of plaster, which hid the delicate facets of the older work, and, when the coarse carving and moulding of Ramesses II was treated with its coat of gesso, the presence of earlier masonry was entirely hidden, and no one would suspect that the imposing columns were composed of scraps eked out with plaster which was sometimes over three inches thick.

Though the word *restoration* in Egyptian actually meant, more often than not, the re-use of the blocks of the older building,[1] repairs were, at

[1] An interesting little point arises in this connexion which might make posterity more lenient in its criticism of kings who apparently usurped the buildings of their ancestors under the pretext of

times, actually carried out in a monument. An interesting example of ancient repairs to columns can be seen in the Great Hall at Karnak. At the east end the faulty parts of the columns have been cut out and filled with a patchwork of smaller blocks (Fig. 166). Here they have not been dressed to the shape of the column, possibly because some other masonry was intended to butt up against them. In the central aisle, however, the repairs have been very carefully dressed (Fig. 167). They probably date to Ptolemaic times, or perhaps slightly earlier.

restoration. The ancient masons, having no lifting tackle (Chapter VIII), could not take down the blocks of a building, if they were of any size, without defacing the scenes and sculptures. On the other hand, it must be admitted that the kings never seem to have ordered that an exact replica of the older scenes be re-sculptured—which the ancient artists were well capable of carrying out. To inscribe a line to the effect that the building had been erected by the ancestor and restored by the reigning king 'after it had been found in a ruinous condition' seems to have been considered ample. Too often, or so it appears, even this was omitted.

Fig. 166. East end of the Great Hall at Karnak, showing ancient repairs to the columns

Fig. 167. Part of a column in the central aisle of the Great Hall at Karnak, showing ancient repairs. Though the patches have been dressed, no sculpturing has been carried out with the exception of the completion of the cartouche of Ramesses II

Fig. 168. Part of the south colonnade of the court of Sheshonq I at Karnak, showing short architrave and patch

Fig. 169. Roof-slab (right) packed up to compensate for an architrave of insufficient depth. Ramesseum, Thebes

XIII

ARCHITRAVES: ROOFS: PROVISION AGAINST RAIN

BEFORE the great sandstone quarries, such as those of Gebel Silsila, were exploited on a large scale, the only means the Egyptians had of roofing a chamber of any size was to have supports for the architrave not more than three yards apart, or else to make use of granite. The enormous labour involved in dressing granite surfaces seems to have deterred the Egyptians from employing it for roofs as a general rule, though in the Old Kingdom, where time and expenditure seem to have been no object, granite is freely used for roofing, as in the Great Pyramid and 'The Temple of the Sphinx'.

When sandstone came into general use, gaps of eight yards or more could be bridged, and this greatly affected the size of the buildings and their proportions. The Egyptians, naturally enough, spanned the largest gaps possible with their materials and, at times, loaded them far beyond what we should consider the margin of safety.

The architraves were placed on the columns with considerable ingenuity, and were often dovetailed together and prevented from slipping while under construction by cramps and peg-dowels (Figs. 121 & 122). Where the Egyptian was unsound in his building-practice was that the bearing-surfaces which he provided for his architraves were often totally insufficient. Though, in the best work, the architrave goes as far back on to the wall as practicable, in too many cases it merely rests on a few inches of bearing-surface, which, if it is sheared, necessarily brings down the architrave with it. Having expended enormous labour on constructing the walls and columns of a building, the architects frequently made use of blocks which were not of the requisite size. For example, in the colonnade in the court of Sheshonq I at Karnak (Fig. 168), one of the architraves is at least eight inches too short, and it has been placed so that it rests on the edge of the abacus of its column, the space between it and the next architrave being filled with a badly cut patch. In the Ramesseum one of a pair of architraves was not of the necessary depth. Instead of obtaining another, small pieces of stone were placed on it to bring it up to the depth of its fellow (Fig. 169). When the packing was covered with plaster and gesso, and painted, no trace of the defective construction would be apparent. It must be admitted, however, that many architraves having these defects have been equal to their load throughout the ages, when others, laid on sounder principles, have failed.

Three methods were in general use by the Egyptians in forming the joint between two architraves on a column. When the two were in a line, a simple straight joint was always used (Fig. 121). When two were required at right angles, the joint ran from the inside of the bend at an angle of 45° until a point either on the axes of the architraves or slightly beyond them[1] (Fig. 170) was reached, the joint then passing out at right angles

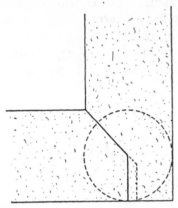

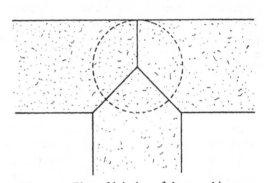

Fig. 170. Plan of jointing of two architraves at right angles; New Kingdom. The dotted line shows the form found in the Vth dynasty temples at Abusîr.

Fig. 171. Plan of jointing of three architraves.

to the line of the architraves. A T-joint between three architraves was made by cutting the end of one to a blunt point to fit into a corresponding V-shaped recess in the other two (Fig. 171).[2]

Though the placing of two architraves side by side—a very common practice—is good construction, saving a great deal of trouble in the matter of laying them and being almost as strong as one of a single block, the Egyptians at times committed the grave fault of forming their architrave of two blocks, one resting upon the other. Such architraves can be seen here and there in the Great Hall at Karnak, sometimes consisting of four

[1] In the figure, the dotted line shows the form of joint found in the Vth dynasty temples at Abusîr (see BORCHARDT, *Das Grabdenkmal des Königs Sahurē*, i, p. 45).

[2] For examples of the arrangement of the archi-traves in 'The Temple of the Sphinx', the temple of Seti I at Abydos, and the temples of Luxor and Karnak, see JÉQUIER, *Les Éléments de l'Architecture Égyptienne*, pp. 281–4.

separate blocks (Fig. 172). It is to be doubted whether the architects realized the great difference in rigidity between two beams placed one upon the other and a single beam of the same height. That the Egyptians must have been alive to the failure of their architraves is certain, but there are very few instances of repairs having been undertaken in those which threatened to fall. In one of the architraves of the Great Hall at Karnak, however, where the lower part has at some time shelled off, holes were cut in the abaci of the columns for the insertion of small beams (Fig. 173). The beams perished long ago, but the architrave still held together as a 'flat arch'. The date of the repair is unknown.

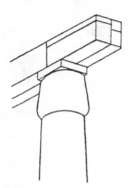

Fig. 172. Composite architrave in the Great Hall at Karnak. (After Jéquier, *Les Éléments de l'Architecture Égyptienne*, Fig. 184.)

In the limestone chambers of the Old Kingdom roof-beams were narrow and deep, and were frequently rounded below to imitate logs (p. 7). In the New Kingdom flat roof-slabs were universally used for roofing; these, though relatively less rigid than blocks laid on edge, were much more economical and involved a less number of joints through which the rain might enter.

The roof-slabs have, generally, suffered more than any other parts of the temples. The reason is obvious; they were in a condition for which the material is most unsuited. Stone will stand a great deal of compression but very little tension. The under-surface of a roof-slab is in tension, while the upper part is in compression. Further, the upper surface is exposed to extremes of temperature, while the lower surface is not. All these factors tended towards disintegration, and the Egyptians' habit of frequently leaving an insufficient bearing-surface for their roof-slabs (Fig. 174) made matters worse. Even in such a relatively modern construction as Edfu Temple, which was built when a tendency towards solid foundations was manifest, the slabs rest on an insignificant angle of wall. Since earthworks and the filling up of the temple were necessary to replace a roof-slab or an architrave, it is astonishing that the Egyptians did not pay greater attention to the solidity of their roofs.

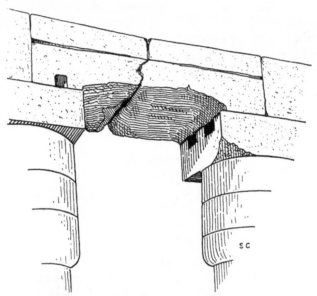

Fig. 173. Holes in the abacus of a column in the Great Hall at Karnak, in which beams appear to have engaged to support the lower part of the architrave, which has now shelled off. The date of the repair is unknown.

In contrast to his lack of care for foundations, the Egyptian took great pains in dealing with the problem of rain. His paints were fairly waterproof (p. 200), and the infrequent rain against the painted surface on the outside of a temple would do no very great harm. Inside it was far otherwise; rain flowing on the roof collects all the mud and dust which have become deposited on it, and the effect of water, with mud in suspension, pouring down a painted surface can well be imagined, especially when it has dried on. In temples, therefore, many more or less ingenious means were employed whereby the joints between successive roof-slabs were made rainproof, and the water led off towards spouts which threw it clear of the temple walls.

Fig. 174. Roof-slab resting on the edge of an architrave in the Festival Temple of Tuthmosis III at Karnak

While in the Old Kingdom temples the water was prevented from entering between the roof-slabs by the mortar between the vertical joints only
(though probably they had a thick layer of mortar above as well), by the
New Kingdom, if not before, it had become usual to cut a square channel,
half in one slab and half in the other, and to fill this with a fillet of stone.
This is the method used in the temple of Seti I at El-Qurna, in the
Ramesseum, in the temple of Ramesses III at Karnak, and in many others
which have still retained their roof-slabs more or less intact. An efficient
variation of this was to leave a ridge on either side of the channel above the

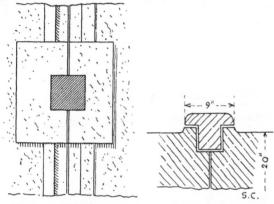
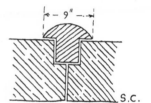

Fig. 175. Joint between two roof-slabs in the temple
of Seti I at Abydos. In the plan, a light-opening is
shown with the joint-trough on each side. In the section, the joint-trough is filled with the stone roll which
rendered the joint watertight.

Fig. 176. Section of a stone
roll filling the joint-troughs
between the roof-slabs in
the temple of Ramesses III
at Medînet Habu.

straight joint between the slabs (Fig. 175), and to fill the channel with a
long piece of stone, rounded at the top, forming, as it were, a roll, which
threw off the rain from the straight joints without depending on the mortar
for making the joint waterproof. In the temple of Seti I at Abydos, where
this method of forming a rainproof joint was used, the joint-channels stop
short of the front face of the temple, possibly to prevent the projecting roll
being seen. In the temple of Ramesses III at Medînet Habu the rolls are
of less height than at Abydos, and there is no ridge on either side of the
joint-trough (Fig. 176). It will be seen that this is less efficient than that
previously described, but it was less in the way if processions were to pass
over the roof. On the eastern roof of the first court of this temple the rolls,
which are nearly all perfect, generally consist of four stones, each about
fourteen feet in length, between each pair of roof-slabs.

In the temple of Ramesses III at Karnak, and in several other structures
of the New Kingdom, the joint-troughs are filled with a square-sectioned
strip of stone which does not rise above the level of the slabs. The only
advantage of this method of constructing a joint is that the distance through
which the rain has to penetrate in order to get through the mortar is longer
than if there were a straight joint between the slabs. It is not impossible
that, in many cases, the projecting rolls have been smoothed off the roof
in later times.

In order to hasten the flow of water from off the roof-slabs, these were
often cut with a slope on them in such a manner that a minimum amount

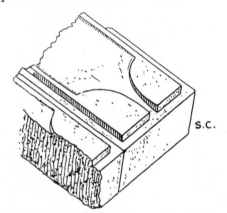

Fig. 177. Roof-slab cut to direct the flow
of rain-water. Temple of Seti I; El-
Qurna.

of stone was removed and that the water was discharged at any required
point from it (Figs. 177 & 178). The discharge-points are not always at
the end of the block, but are found also at the sides and even in the middle,
depending on the position of the main gutter by which the water from the
slabs was led off to the spouts. The arrangement of the gutters varies in
different temples, the temples of Seti I at El-Qurna (Fig. 179) and
Abydos (Fig. 180) being good examples for study. In the temple of
Abydos, where a joint-trough crosses a gutter, the stone is cut to form a
square hole of about eight inches side, which was, no doubt, filled with a
stone through which the gutter was continued. The rolls naturally stop
short of such intersections.

Another method of dealing with rain was to have the whole roof dressed
to a slope to lead the water in the desired direction. This method is generally

Fig. 178. Roof-slabs over the centre aisle in the temple of Seti I at El-Qurna, arranged to permit the rain-water to run off on to those of the side-aisles

used in conjunction with a mosaic of small blocks laid above the roof-slabs
(Fig. 181). Though more usual in the late temples, such as Dendera, the

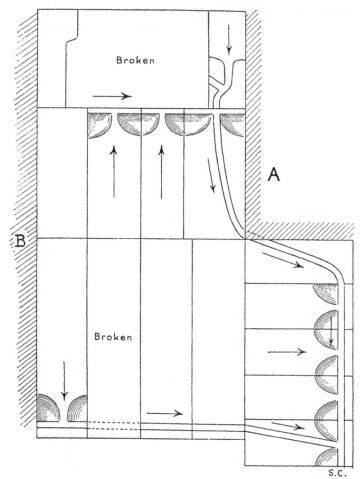

Fig. 179. Sketch-plan of part of the lower roof of the temple of Seti I
at El-Qurna, showing direction of the flow of water off the roof-slabs
and along the gutters. (Higher roofs at *A* and *B*.)

system of covering the roof-slabs with a mosaic of polygonal stones is known
quite early in the New Kingdom. In the small temple of Tuthmosis III at
Medînet Habu part of the roof has the slabs covered with an upper coating
of such stones. At the edge of the mosaic there is a slight ledge to prevent
the water from overflowing, and to lead it to the place where it could be

conveniently discharged. In another part of the roof the projecting roll (Fig. 176) was cut away so that a mosaic could be laid over it.

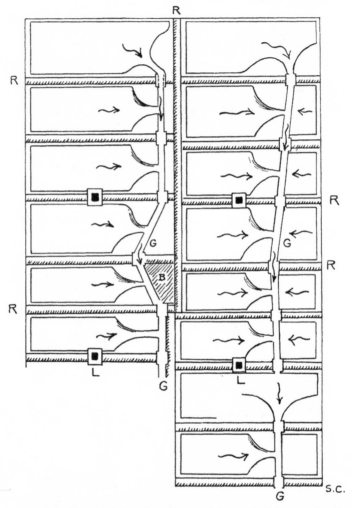

Fig. 180. Sketch of the plan of part of the roof of the temple of Seti I at Abydos, showing the direction of flow of water off each roof-slab into the gutters (G). B, roof destroyed. R, rolls filling up joint-troughs between the slabs. L, light or ventilation openings.

In some temples, notably the Festival Hall of Tuthmosis III at Karnak, no drainage system is apparent. The blocks have no joint-troughs between them, and are not cut to facilitate the flow of water over them. Possibly there was a thick layer of mortar or plaster over the whole roof.

In the Temple of Isis at Philae the roof was covered with a mosaic of smaller blocks, but these have largely disappeared. The water was led off the roof by a careful arrangement of the slope of the blocks as a whole, and not by leading the rain off individual blocks into specially cut channels.

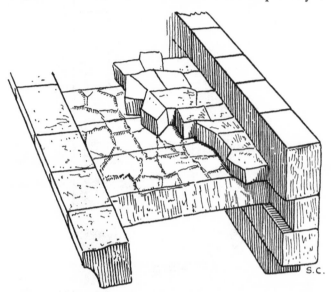

Fig. 181. Pavement of polygonal blocks laid over roof-slabs. Temple of Dendera. Roman date.

The water in this temple is conducted to spouts, which discharge it into the small open courts below, whence it is led outside the temple.

The Mammisi or 'Birth-house' at Philae had a somewhat similar system of roof-drainage to that of the large temple. The water from the upper roof (Fig. 182) was made to flow to two water-spouts, the flow being accelerated, near the main spout, by channels cut in the roof-slabs. On this roof there appears to have been no upper pavement. The lower roof (D) had its own spouts and discharged its water on to the colonnade below. An interesting feature of the roof of the colonnade is to be noticed in some of its roof-slabs, which are set almost radially and are of very different shapes. This arrangement of roof-slabs can also be seen in the roofs of the XXVth dynasty shrines at Medînet Habu.

Water-spouts show very little variation of form in Egyptian architecture,

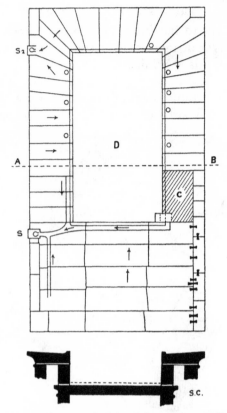

Fig. 182. Drainage system of the roof of the 'Mammisi' at Philae. Below is a section through *AB*, the depression in the top of the cornice being exaggerated. *C*, roof broken away. *D*, lower roof, covered by a pavement of polygonal blocks. *S* and *S*2, water-spouts. The arrows show the direction of the flow of water.

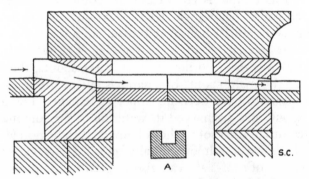

Fig. 183. Section of the upper part of a colonnade in the Temple of Dakka, showing how a water-channel was carried under the roof of the colonnade to drain the lower roof behind it. *A*, section of the channel. Roman date.

Fig. 184. Drip-channel in the outer wall of the temple of
Ramesses II at Abydos

the forms being either a lion's head, which is known from the Vth dynasty
to Roman times, or a plain projecting spout-block in which is cut a rectangular channel. In the temple of Dakka in Nubia an exceptional method
of draining the roof can be seen. Here it was desired to carry the water from
a roof without permitting it to flow on to the floor of a colonnade with a
higher roof which was built around it; a channel was accordingly constructed which passed, bridge-wise, just below the roof of the colonnade.
The channel, where it spans the gap between the two walls, has, curiously
enough, a joint in the middle (Fig. 183).

In the temple of Ramesses II at Abydos the wall on the outside of the
temple has a vertical recess of semicircular section cut in it (Fig. 184). The
sculptures pass inside this recess, showing that it is contemporary with the
masonry. Its function may have been to catch any drips from a water-
spout, and prevent them from spreading over the painted surface of the wall.

XIV

DOORS AND DOORWAYS

THE Egyptians gave considerable importance to the doorways in the temples and tomb-chapels. Whilst northern races had of necessity to shut out cold and intemperate weather, which tended to keep the doors of comparatively modest dimensions, the southern races could use them not only as an entrance, but as the main source of light.

The doorways between the pylon towers were often of great height—sometimes forty feet or more—and a type of door-frame was invented, all in stone, which, while maintaining the majesty of the doorway, did not require that the doors themselves should be of impossibly large size.

The great doors of the temples have naturally disappeared long ago, and ancient records furnish no evidence concerning their structure beyond relating how they were overlaid with gold or electrum. Information has to be gathered from the representations of doors in tomb-chapels, and from the doors of modest dimensions which chance has preserved.

Among the wooden doors which have been preserved from dynastic times may be cited the door found in the mastaba of Neferma'ēt of the IVth dynasty from Meydûm,[1] the solid door from the Vth dynasty mastaba of Kaemhesit at Saqqâra[2] (Fig. 185), and one from El-Lahûn (Fig. 186) which, from an almost erased scene on it containing cartouches, appears to be of the time of King Osorkon I of the XXIInd dynasty.[3]

The door of Neferma'ēt was never hinged, at any rate when placed in the tomb. It measured 28·75 inches in width and was composed of two thick planks fastened together by countersunk cross-pieces. On either side of it stood two wooden jambs, and to secure it in place a great wooden beam, 16 inches thick, was let down on to it.

The fine wooden door from the mastaba of Kaemhesit is of one piece, and is in perfect condition except for warping. The sculptor's name, Ithu, is carved on it behind the leg of the noble's figure, and forms one of the rare instances where a craftsman has been permitted to put his name to his work.[4] It is provided on the back with a series of seven horizontal battens.

---

[1] Published in PETRIE, *The Labyrinth, Gerzeh and Mazghuneh*, p. 25 and Pl. XVI.
[2] Now in Cairo Museum, *Journal d'entrée*, No. 47749.
[3] In Cairo Museum; temporary register, No. 20.5.24.4.
[4] Another instance is seen in the statue of Zoser dedicated by Imhōtep and 'signed' by the sculptor. (See GUNN, *Annales du Service*, xxvi, p. 177.)

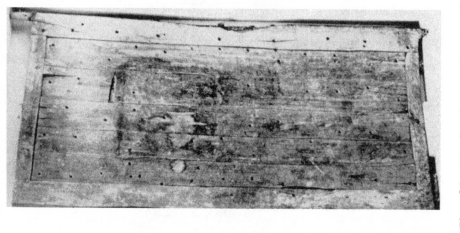

Fig. 186. Wooden door of Osorkon I from El-Lahûn. Cairo Museum. (Photograph by M. G. Oropesa)

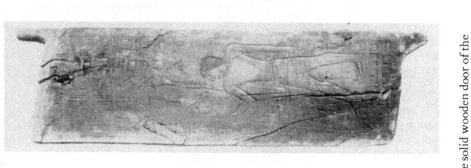

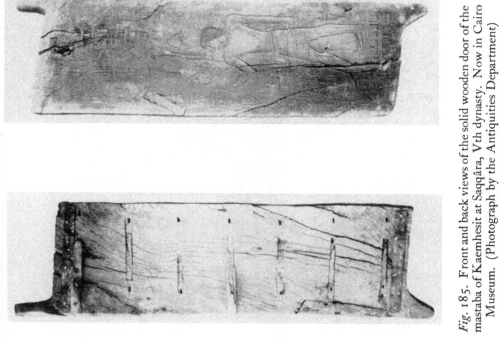

Fig. 185. Front and back views of the solid wooden door of the mastaba of Kaemhesit at Saqqâra, Vth dynasty. Now in Cairo Museum. (Photograph by the Antiquities Department)

Fig. 187. 'False door' in the mastaba of Sesheshet at Saqqâra, showing imitation of the two leaves, bolt and lower pivots. Vth dynasty

Fig. 188. Upper and lower door-pivots of bronze. Scale $\frac{1}{8}$, Cairo Museum

These are not sufficiently stout to have been of any use in preventing warping, and may well have been placed there to maintain the tradition of a door constructed of planks.

Doors are frequently represented in stone. In the masonry of Zoser some of the shrines have imitations of open doors which are part of the masonry of the building. One can be seen on the extreme left of Fig. 3.[1] The panelled boundary wall of the Step Pyramid also has the representation of a large double door on the middle of the south side in the masonry, but beyond the vertical line showing that they were double, and the bottom pivots, no other details are given (Fig. 91, B). The most frequent type of doors represented in stone is what are now called 'false doors', through which the spirit of the deceased was supposed to come from his tomb-chamber into his chapel to receive the offerings made for him. They are often sculptured and painted with all the details of a real door. Good examples can be seen in the Cairo Museum and in the Vth dynasty mastaba of Mereruka (Mera) and his wife Sesheshet at Saqqâra (Fig. 187). Here the doors are clearly represented as being formed of a series of planks held together by horizontal battens. It is not known whether the great temple doors, some of which must have weighed three or four tons, were of this form or not.

The Egyptians did not make use of hinges of modern form for their doors, though the principle of the hinge[2] was known to them and used for the lids of boxes. A hinge with one plate attached to the door and the other to the door-frame is at best weak, since a heavy door will sink, in the course of time, by its own weight, and is apt to bend it out of shape. In the Egyptian method of hanging a door no frame was necessary; their doors moved on pivots, one rising upwards and fitting into a socket in the lintel and the other going downwards and revolving in a recess in the sill or in a hard stone block. The pivots of the wooden doors already described are all in one piece with the woodwork of the door itself; in others, and certainly in the great temple doors, they were of copper or bronze. In all doors and representations of them, the top and bottom pivots have forms which hardly varied throughout the whole course of Egyptian history (Fig. 188).

A large temple door was hung by passing up the top pivot into the circular socket in the lintel and dropping it so that the bottom pivot engaged

[1] See also FIRTH, *Annales du Service*, xxiv, Pl. 1; No. 2.
[2] See CARTER and MACE, *The Tomb of Tut·ankh·* *amen*, Pl. LVII, where hinges of modern form are shown on an ivory casket.

in the socket in the sill. In order to do this, the top pivot had to be made sufficiently long to engage in the socket after the door had been let down. The result of this method of hanging a door was that the leaves could not be a good fit at the top, since a considerable gap was left between them and the lintel. This was lessened by cutting a groove, sometimes sloping, leading down to the bottom of the lower socket, which was filled in with a strip of stone after the door had been hung. By this means, when the upper pivot had been pushed into place, the lower pivot could be passed along the groove into its socket, and very little letting down of the door was necessary. Sometimes the lower part of the upper socket was of wood. This could, if necessary, be raised while already engaged in the top pivot of the door, and fixed by wedges when the door was in its place. The

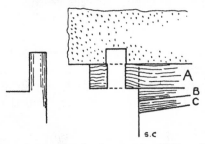

Fig. 189. Upper door-socket in the temple of Kôm Ombo, showing manner in which the upper pivot passes through a block of wood (*A*) held in position by wooden wedges (*B C*).

remains of such a socket with part of the wood still in place are found in the temple of Kôm Ombo (Fig. 189), and recesses in the top of the jambs, apparently for a similar purpose, exist in other temples. At the temple of Seti I at Abydos one of the recesses still contains the remains of a palm-wood socket. A good example of a groove for the passing in of the lower pivot can also be seen in the temple of Kôm Ombo (Fig. 190), the length of the groove being 2 feet 3 inches and the depth where the pivot rested 4½ inches.

In temples the doors were almost always hung so that they opened back into recesses in the sides of the doorway.

The stone frames forming the doorways between the towers of a pylon show several peculiarities. In the unfinished pylon (Fig. 87) at Karnak it will be seen that the doorway, which has been dressed smooth, is not bonded into the masonry of the towers. Pylon towers, even when constructed with stairways and chambers within them, are very heavy pieces of masonry and would tend to sink somewhat by their own weight, particularly when one considers how poor were Egyptian foundations in most cases. The more free such masses were to move, the less likely would they be to drag down the door-jambs with them. It is probable that the construction of the doorway with straight joints between it and the pylon towers had, for its intention, the avoidance of this difficulty. If, however, the pylons of the different dynasties are examined, it is noticed that the doorways are bonded into the pylon towers as often as not, and no rule seems to be observed in this matter. For instance, in the Tuthmosid pylon (No. VIII) at Karnak, in Luxor Temple, in the temple of Khonsu, and in Edfu Temple the doorway is bonded into the masonry of the pylon towers, while in the temple of Medînet Habu (Ramesses III), the unfinished pylon

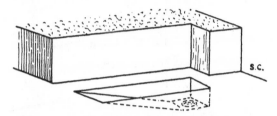

Fig. 190. Lower socket of a doorway at the temple of Kôm Ombo, showing groove in the sill to enable the door-pivot to be passed into its socket.

at Karnak (No. I), the Ethiopian pylon at Medînet Habu,[1] and others the doorway merely butts on to the remainder of the masonry of the pylon. In cases where the doorway forms a separate entity the edge of the cornice usually passes into the masonry of the towers, with the result that the settling of the masonry of the towers often breaks them off. In the Ethiopian pylon at Medînet Habu this risk is obviated by the doorway being formed, as it were, in a square frame of masonry, which leaves about two feet between the extreme edge of the cornice and the tower. This frame of masonry butts on to that of the pylon towers. The effect of this method of construction has been to preserve the cornice intact.

In the Roman doorway at Dendera it seems as if the intention had origi-nally been to construct a pylon, which never materialized, and it has been inferred from this that it was the Egyptian custom to construct the door-ways of pylons first and the towers afterwards. In this doorway the faces which would have abutted on to the towers are not faced except for about

[1] JÉQUIER, *Les Temples Ptolémaïques et Romains*, Pl. 12.

a foot in from the front face; the remainder is left in a rough state in precisely the same manner as the gateways which engaged in the brick girdle-walls (Fig. 191). Though in isolated cases a doorway may have been constructed before the towers, it is very unlikely that it was the rule, even

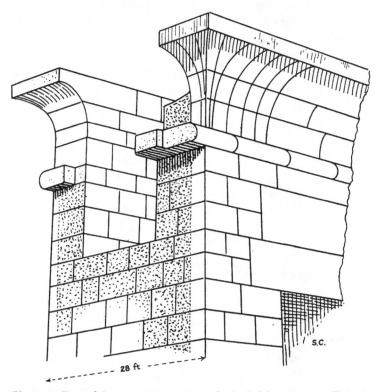

Fig. 191. Part of the great stone gateway in the brick temenos wall north of the temple of Monthu at Karnak. The surface in contact with the brick was left rough. The wall must have been at least as high as the top of the cornice, which is about 50 feet.

when the doorway and the towers are not interbonded. Such a method of construction would be wasteful in the extreme, since great embankments (p. 91) would have to be made and removed for each piece of work and no advantage would be gained. In the unfinished pylon at Karnak, though the door-frame is separate from the towers, the courses of the door-frame and pylon are at the same height throughout, in contrast to those in the temple of Medînet Habu. Since the method of laying a block (p. 98)

Fig. 192. Gateway north of the temple of Monthu at Karnak which engaged in the great temenos wall which once enclosed the temple. Ptolemaic date

Fig. 193. Gateway of Ptolemy II, south of the temple of Khonsu at Karnak. The brick temenos wall round the main group of temples passed from here eastwards to the southern pylon of Haremhab

involved the dressing of the top of the course after it was in position, it is very likely that the whole was constructed course by course. Nor need it be assumed that the Medînet Habu doorway was ever, during construction, at a great difference of level from the remainder of the pylon masonry. In this case the usual Egyptian method was followed of using blocks of irregular height, while in the unfinished pylon of Karnak the intention was to keep the bedding joints of the courses regular.

The stone gateways constructed in the great brick temenos walls of Karnak are too well known to require any detailed description (Figs. 191–3). Gateways of this type appear shortly before Ptolemaic times. The sides, against which the bricks lay, are left rough, and it is clear from the break in the cornice (Fig. 191) that the brick walls were as high as the masonry, which is about fifty feet. By the unfinished pylon, and in many places round the group of temples, the wall can still be traced (Fig. 253); it is, in places, nearly thirty feet thick. By the gateways it has, however, almost entirely disappeared.

In temples many doorways are of quite simple form, a lintel being laid across the regular masonry of the wall. Often the jambs and lintel are of hard rock such as granite. This form of doorway offers no constructional problems of any interest.

The method of hanging a door in ancient Egypt by means of vertical pivots gave rise to a peculiar form of door-frame for which we seem to have no term in architecture. There is no true lintel, but the top pivots engage in masonry projecting from the jambs, which was almost invariably decorated with the cornice moulding on the front and back and on the inside. This type of frame is known as early as the XVIIIth dynasty, where it can be seen represented in the plans on the tomb-walls at El-'Amârna (Fig. 55). When a door was to be fitted to a hypostyle hall whose front face consisted of columns with screen walls between them (Fig. 205), such a form of doorway was about the only one which would be tolerable from the point of view of appearance, and it is the form very commonly used in late temples. This type of door was even used in stone in the middle of brick walls, which must have given an effect the reverse of pleasing.

The sills of the great doors of the temples often stand as much as a foot above the level of the remainder of the pavement, and are usually of hard rock. The door shut against these sills. The reason may have been to prevent water running into the inside of the temple.

The doors of pyramids have formed the subject of a considerable amount

of speculation. Strabo, referring to the Great Pyramid, relates that there is[1]
'a stone that may be taken out, which being raised up, there is a sloping
passage'. This has been taken to mean that there was a stone, hinged at the
top of each side, which opened flap-wise. Some support is given to this
idea by an examination of the south pyramid of Dahshûr,[2] where there are
recesses near the top of each side of the entrance which suggest that a flap-
door may have swung in them;[3] but the entrance to the Third Pyramid at
Gîza (Fig. 99) shows no such recesses at all, and it cannot be believed that,
once kings were buried in their pyramids, they were in any way open to
the priests or to the public; their temples were the places where the offerings and the prayers were made. In the south pyramid at Dahshûr there
are two entrance passages descending into two separate burial chambers
with corbelled roofs. These chambers are at different levels, but one now
breaks into the other.[4] Though one entrance is now open, the other is

Fig. 194. Bronze door-bolt from Memphis. Scale ⅓. Cairo Museum.

completely blocked up with stone the whole way down the entrance passage. This must surely have been the manner of closing all the pyramids
when they had received the body of the king. All the pyramids were
apparently robbed in that period following the fall of the Old Kingdom
known as the First Intermediate, though the worship of the king in the
vicinity of many of the pyramids was carried out until quite late times.[5]
By Strabo's time it is not impossible that the entrance to the Great Pyramid
may have been fitted with some such door as he describes, though there are
no traces of one to-day, as the casing has disappeared.

The portcullis, being, as it were, a form of door, merits a brief notice. It
was used to close entrances of underground passages from very early times.
The great mastaba of the IIIrd dynasty at Beit Khallâf,[6] believed by some
to be that of King Zoser, has its descending passage barred by five great
portcullis blocks of stone engaging in masonry, and others can be studied

---

[1] Ed. Bohn, iii. 249.
[2] The date of this pyramid is unknown, and it may
possibly come between those of Zoser and Sneferu.
It has the peculiarity of its casing abruptly changing
its slope. At the time of writing, excavations are
being carried out on this monument.
[3] See Petrie, *The Pyramids and Temples of Gizeh*

(new ed. 1885).
[4] Perring, *The Pyramids of Giza*, Pl. VII.
[5] Twenty-one priests of Khufu's pyramid are
known, extending from the IVth to the XXVIth
dynasties. See Petrie, *History*, 1923, p. 60.
[6] Garstang, *Mahâsna and Bêt Khallâf*, Pls. VI
and VII.

in the pyramids of Gîza, Saqqâra, and Abusîr. As a protection to the tomb chambers, once entrance had been effected within the pyramid or mastaba, portcullises proved of little value, since the tomb robbers rarely attempted to interfere with them, but merely cut a passage round or under them.

The common form of Egyptian bolt for a two-leaf door is a piece of metal or wood flat on one side and rounded on the other which slides in two staples on one leaf and in one or two on the other (Fig. 194). Actual bolts in bronze and wood have been found in town sites, and examples are frequent on the doors of small shrines. An examination of the door-frames in the temples shows clearly not only that vertical bolts were used which engaged into the sill and the lintel, and bolts for single-leaf doors engaging in the jamb, but that the doors could be barred from inside, recesses into which the bar was slipped being not uncommon in temple doors. The form of the vertical bolts is not known, and that of the bolts working into the jamb can only be determined by deduction,[1] since no actual examples of this sort have yet been found.

[1] For reconstructions of the bolts of the doors of the temples at Abusîr, see BORCHARDT, *Das Grabdenk-* *mal des Königs Sahurē'*, Fig. 34, p. 38, and Fig. 72, p. 59.

XV

WINDOWS AND VENTILATION OPENINGS

IN Egypt, with its penetrating sunshine, windows played a comparatively small part in architecture until the New Kingdom, when the
clerestory and screen walls became popular. The true source of light was
the door.

A very primitive form of window, which was used throughout the history
of Egyptian architecture, consists of a narrow slit, usually opening into the

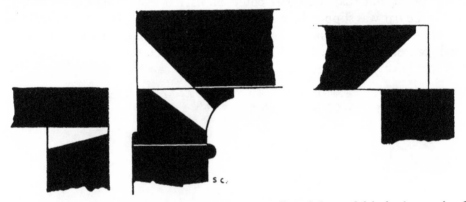

Figs. 195–7. Sections of light-openings between the walls and the roof-slabs in the temple of
Tuthmosis III at Medînet Habu.

temple just under the join of the roof and the wall. These slits were far too
high to see through, and too narrow to enable anyone outside on the roof
to look inside. Among the earliest examples are those in the valley temple
of the Second Pyramid ('The Temple of the Sphinx') at Gîza. The exact
direction which the slit windows take through the wall and roof-slabs seems
to have depended on convenience; some are cut entirely from the wall,
some from the roof-slabs, while others pass through both; they usually
open out on to the sides of the temple, but occasionally they pass out to
the top of the roof (Figs. 195–201). In the seven parallel sanctuaries of the
temple of Seti I at Abydos they enter the temple in the curve of the false-
arch (Fig. 220). This form of light-opening is very rarely decorated.
In Dendera, however, the sloping sills of many of the slits are carved to
represent the rays of the sun entering through them.

In most of the temples in which the flat stone roofs, or fragments of them, survive, small openings can be seen (Fig. 175) through the roof which

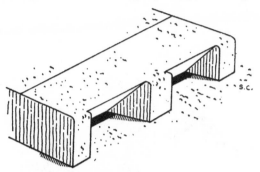

Fig. 198. Light-opening in a roof-slab in the temple of Tuthmosis III at Medînet Habu.

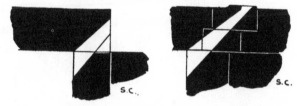

Fig. 199. Section of light-opening at Deir El-Medîna. Ptolemaic date.

Figs. 200, 201. Sections of light-openings in the temple of Opet at Karnak. Ptolemaic date.

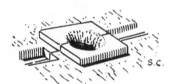

Fig. 202. Ventilation opening between two roof-slabs. Temple of Seti I at Abydos.

would admit a certain amount of light and also serve for ventilation, and considerable ingenuity is often shown to avoid weakening the roof, the holes being pierced at the junction of two roof slabs.

The extraordinary ventilation-channels in the Great Pyramid are too well known to require a detailed description. It suffices to remark that they serve the apartments known as the King's and the Queen's Chambers.

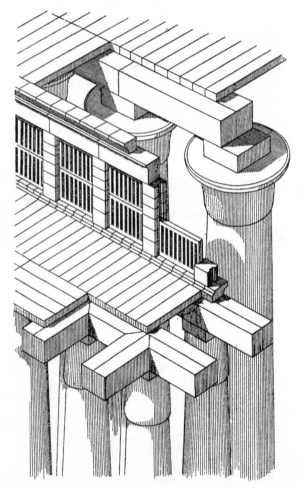

Fig. 203. Clerestory of the Great Hall at Karnak, showing details of the masonry. (Aft\
PERROT & CHIPIEZ, *Histoire de L'Art dans L'Antiquité*, i, p. 615.)

Note.—In this drawing the roof-slabs are, it appears, too narrow and not deep enough.

Fig. 204. Clerestory window of the Great Hall at Karnak.
(Photograph by Gaddis and Seif, Luxor)

Fig. 205. Façade of the temple of Dendera

After running for a short distance horizontally, they suddenly turn upwards at an angle of about 30° for some seventy yards, to come out on the face
of the pyramid. The internal section is square, the channel being cut out
of one stone, while another stone forms the lid. They are not pierced
through blocks laid on horizontal beds, but the blocks in which they are cut
are laid at the same angle as the line of the ventilation passage, as is the case
with the large passages in the pyramid. Where these channels have been
broken away to make a forced passage, their construction can be easily
studied.

The clerestory is not found before the New Kingdom. The remains of
a magnificent example of this period is in the Hypostyle Hall at Karnak
(Figs. 203 & 204). They are also found in the temple of Seti I, in the
Ramesseum at Thebes, and in the temple of Khonsu at Karnak. They
seem to have given way in later times almost entirely to the practice of
using screen walls between the pillars forming the first hypostyle hall of the
temples (Fig. 205). At any rate, in the Ptolemaic and Roman temples,
such as Edfu and Dendera, no clerestories were used. Screen walls, as far
as we know from surviving monuments, antedate the clerestories, a fine
example with very low screen walls being seen in the XVIIIth dynasty
temple at Medînet Habu.

It is not until very late times that windows are found in the outer walls
of the temples. The earliest examples of such windows are seen in the little
temple of Hakor at Karnak, where they are cut in the form of gratings
from a single block of stone.

Ornamental windows seem to have been known from very early times.
In the masonry of Zoser, the side of one of the chapels was decorated with
a row of hieroglyphs known as the *dad*-sign.[1] There is no doubt that these
are a representation of a window, since the space between three or four
signs had actually been cut through, converting the signs into mullions. In
the XIth dynasty, a false door is known, surmounted by a representation of
a grilled window, in which hieroglyphic signs and floral patterns form the
mullions.[2] A similar window is represented on the door of a model kiosk
and garden found in the XIth dynasty tomb of Meketrē' at El-Deir el-
Bahari,[3] and others in the sanctuaries of the temple of Seti I at Abydos.[4]
The palace of Ramesses III, which once stood at Medînet Habu, was furnished with mullioned windows in which divine figures and cartouches as

[1] C. M. FIRTH, *Annales du Service*, xxv, p. 158.
[2] NAVILLE, *The Eleventh Dynasty Temple of Deir el-Bahari*, ii, Pl. XIV.
[3] Now in Cairo Museum (*Journal d'Entrée*, No. 46721).
[4] JÉQUIER, *Les Temples Ramessides et Saïtes*.

Fig. 206. Sandstone window from the palace of Ramesses III at Medînet Habu, now in Cairo Museum. Scale approximately ⅓. (From drawings by Yûsif Effendi Khafâgy.)

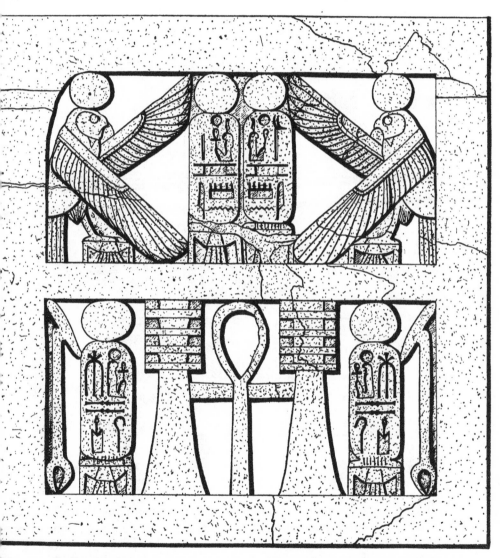

Fig. 207. Sandstone window from the palace of Ramesses III at Medînet Habu, now in Cairo Museum. Scale approximately ⅕. (From drawing by Yûsif Effendi Khafâgy.)

well as hieroglyphs were introduced to form the mullions (Figs. 206 & 207). These are of comparatively coarse work in sandstone. It is strange that,

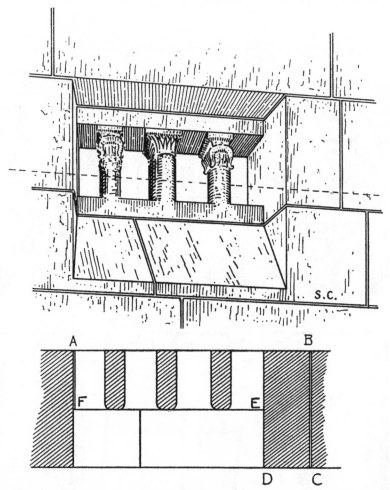

Fig. 208. Window in the temple of Deir el-Medîna at Thebes. The part lettered *A–F* is of one block. Ptolemaic date.

though they were discovered many years ago, no complete publication of them has appeared. Still another window reputed to have come from Dendera, in which pillars surmounted by Hathor-headed capitals form the mullions, is cited in Budge, *Egyptian Sculptures in the British Museum*, pl. xlix. The block forming it is shaped like the façade of a temple and decorated with two winged serpents on the cornice and the architrave, and flanked by serpents, wearing the crowns of Upper and Lower Egypt on a lily and papyrus plant respectively. It is very likely that fancy windows of the forms described were common in Egyptian buildings and houses, but were not generally used in temples.

At Deir el-Medîna there is a very curious ornamental window (Fig. 208) which shows some interesting points in its construction. Its duty was to light a stair leading to the roof. It has three mullions, which present a flat unornamented face to the exterior, but are of semicircular form when seen from within. Two of the mullions are decorated with Horus-headed capitals, while the centre one is carved with leaves. In the plan, the part *A B C D E F* is of one stone, and it is certain that the window was cut through after the blocks had been built into place.

XVI

STAIRS

STAIRS were known in Egypt from the earliest times. In the first
dynasty, a very large brick stairway was found leading down into the
tomb of King Den[1] at Abydos and another into that of his successor 'Az-
iēb.[2] Stairways cut in the rock are known from an equally early period,

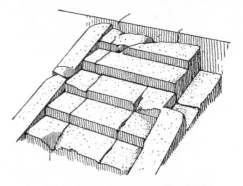

Fig. 209. Stairway in the temple of Edfu.
(After a photograph in JÉQUIER, *Les Temples
Ptolémaïques et Romains*, Pl. XXXII.)

many examples having been found leading down into the early dynastic
tombs at El-Lahûn.[3]

The earliest stairway yet known which is constructed of laid blocks is
one leading up to the roof of one of the small chapels which are believed to
have formed the 'heb-sed' or festival temple of King Zoser at Saqqâra (Fig.
210). It is free-standing, and shows peculiarities not encountered elsewhere. Each step is formed of a separate block which engages in a small
recess cut into the block below it. The angle of the 'riser' or front of the
step is not vertical, but nearly at right angles to the surface or 'tread'. The
treads have a very steep angle of slope and the stairway resembles a plain
dromos more than any other example of later times. Here, however, in
contrast to later stairways, each step has been separately constructed. In
later times it was almost always the custom to lay the blocks first and to cut

[1] PETRIE, *Royal Tombs*, ii, Pl. LVI A, Nos. 3 and 4. [3] PETRIE, *Lahun*, ii, p. 22 and Pls. XLI, XLII.
[2] *Ibid.* i, Pl. LXVI, 2.

Fig. 210. Stairway leading up to the roof of the festival temple of King Zoser at Saqqâra. IIIrd dynasty. (Photograph supplied by Cecil Firth, Esq.)

Fig. 211. Remains of a stairway in the pylon of Tuthmosis III (Pylon VII) at Karnak

Fig. 212. Remains of a stairway in the pylon of Amenophis III at Karnak

Fig. 213. Line of roof-slabs covering the stair-passage in the pylon of the Ramesseum

the steps in them afterwards. Sloping treads are very common in Egyptian buildings, but examples are known both in the Old and the Middle Kingdoms in which the steps are level and the risers vertical.

The stairways leading up to the doors of temples, when each successive hall is at a higher level than the last, are clearly developments of the dromos. In these, the riser is very low and the tread sloping. At times they were flanked by a plain border at an even slope from the top to the bottom, and at other times stone was left to form a low parapet, generally with a rounded top, on either side of the steps. From the total lack of connexion between the steps and the rising and bedding joints of the blocks from which they are built, it is certain that the steps were cut after the masonry was laid (Fig. 209).

In the XVIIIth dynasty temple at El-Deir el-Bahari, a stairway leading up to a platform or altar presents some peculiar features. It appears that an evenly sloping way was first constructed and the surface dressed, and that, afterwards, a new series of blocks was laid on this surface, in which stairs were cut. The advantages of such a system of construction, if any, are not manifest. It may possibly have been a change of plan on the part of the architect.

Most of the larger pylons were provided with internal stairways, which, as a rule, passed from a small doorway in the end of one of the towers up to the level of the door-lintel, whence other stairways led up to the roof of each tower. During the construction of certain pylons, such, for example, as the unfinished pylon at Karnak (Fig. 87), a corresponding sloping passage led from the end of the other tower to the lintel of the doorway, which was filled in after the pylon was completed. In such late pylons as that of the temple of Edfu, the stairways, which are winding, are kept open in both towers, and give access to the many chambers which were constructed in them (Fig. 131).

The method of constructing a stairway in a pylon shows two modifications, generally—but not invariably—depending on whether the pylon consists of a solid mass of more or less well-laid blocks, or whether it is but a stone 'box' into which odd blocks of stone were heaped (Fig. 130). In the former type the whole stairway is constructed of blocks laid on horizontal beds, and one block may include not only several steps but part of the side of the passage as well (Fig. 211). In the latter method the blocks from which the steps were to be cut were dressed to a continuous slope, and on this slope were laid the blocks for the walls of the passage (Fig. 212). It is obvious that the former method is conducive to much greater stability in

the masonry of the pylon. In the XVIIIth and XIXth dynasties, both the
forms of stairway are found, but in the Ptolemaic and Roman pylons the
blocks are almost always laid on horizontal beds. In nearly all the stairways
within the pylons the upper surface of the sides of the passage was dressed
to the slope of the stairway, and roof-slabs were simply laid·across from one
to the other, being left rough on the outside (Fig. 213). No precautions
were taken to prevent the roof as a whole from sliding along on its bed. In
the Grand Gallery of the Great Pyramid, however, the lower edge of each
roof-slab engaged in a recess cut into the top of the side-walls (Fig. 228).

In the Royal tombs at Thebes, when the descending passage is steep, it
is cut into steps with a plain sloping way left in the middle. This was possibly to prevent the sarcophagus damaging the steps when it was introduced
into the tomb. In the Grand Gallery, leading up to the King's Chamber
of the Great Pyramid, the steps are in the middle, two side slides being left
next to the walls. No satisfactory explanation has yet been forthcoming for
these slides nor for the deep recesses cut at intervals along their length.

It must not be supposed that the Egyptians never built up stairways
whose steps consisted of more or less separate blocks. In certain of the
storehouses in the Vth dynasty temples at Abusîr,[1] stairs are found, built
up against the walls leading up to the roof or upper story, which are constructed of comparatively small blocks. In one case[2] the guide-line indicating the slope of the stairway can still be seen painted on the wall against
which the stairway was to be built.

The modern form of stair, where each step is a cantilever engaging in the
wall, is not known in Egyptian stonework. With wooden beams, however,
it may well have been the common form of house stairway.[3]

[1] BORCHARDT, *Das Grabdenkmal des Königs*   [2] BORCHARDT, *Das Re-Heiligtum des Königs Newo-*
Neferirkerēʿ, pp. 33, 34.   *serrēʿ*, p. 23.   [3] *Ancient Egypt*, 1916, p. 171.

XVII

ARCHES AND RELIEVING ARCHES

THOUGH brick arches are known as far back as the first dynasty, none in stone has yet been found earlier than the Middle Kingdom. The stone arch of modern type, where the voussoirs are held in place and support one another by friction, hardly exists in the Egyptian monuments. A false arch, cut from already laid blocks in a corbelled or pointed roof, is the only form used.

Brick arches are of two kinds: those constructed with the ordinary building brick, and those in which a form of brick specially designed for this purpose is employed.

The first type of brick arch dates from the earliest times, but never, to our knowledge, has a span of more than a few feet. In the great mastaba of the IIIrd dynasty at Beit Khallâf, one is found roofing the descending passage. It is described by GARSTANG in *Mahâsna and Bêt Khallâf*, p. 9, as follows: 'The arch, like others of the same period which have been found, was built on the ordinary form of bricks placed edgeways side to side and packed above with pebbles and mud mortar to provide the necessary wedge-form. . . .' Though in modern times such an arch is constructed on a support, called 'centring', which is removed after it is complete, when the span is small, one can be constructed without centring, and it seems—though this is not certain—that none was used in ancient Egypt for brickwork, otherwise large arches of this type would be expected, and, in the small examples, neater work than that actually observed. It suffices, however, to realize that the principle of the true arch must have been known. Arches with the bricks laid lengthways—as stretchers—along the axis of the arch as described in the mastaba at Beit Khallâf, were used at most periods, especially for roofing small tombs, and they are very occasionally found constructed of small, rough blocks of stone. On the outside of the arch, where the bricks gap, the spaces are always filled in with pebbles or sherds.

The second form of brick arch, which was used for spanning very considerable distances and for supporting great weights, was constructed of much thinner bricks than those used for building walls. They measure about $14 \times 7 \times 1\frac{1}{2}$ inches, and were scored heavily with the fingers on one or both of their largest faces while still wet. They were laid on their longest edge in the form of rings (Fig. 214), each ring leaning against the one next to it. Though they occur in the Old Kingdom, the best known arches of this type are those in the Ramesseum (Fig. 215). To construct them, they had to be commenced from a wall at the end of the tunnel. If, as in the case of an archway, there was no wall at the end, one had to be constructed, which was removed when the arch was complete. Pebbles or sherds were inserted to fill in the gaps between the bricks on the outside of the ring in exactly the same manner as for the first type of arch. Having thus con-

Fig. 214. Method of constructing an arch without centring, by means of special bricks. Ramesseum; Thebes.

structed an arch or rings, other rings were laid above it, with the bricks sometimes on the slant and at other times laid upright. In great archways, the rings of special bricks were used as a centring for a succession of rings of ordinary building brick laid as stretchers in the direction of the tunnel (Fig. 216). Though used as centring, the rings of special bricks were not removed when the upper rings were complete. In the great brick arches in the gateways at El-'Asasîf at Thebes, one, with a span of 13 feet, has six rings of brick still remaining, about four having disappeared from inside. Here it cannot be determined how many rings of special arch-bricks were used. In the other archway, whose span is 9 feet 6 inches, one ring of special bricks supports three of ordinary bricks. This is also the case with the archway in the brick pylon of Sheshonq at Dirâ' Abu'l Naga at

Fig. 215. Brick arches, constructed without centring, in the enclosure of the Ramesseum. Date uncertain

Thebes. In the tunnel through the temenos wall of the 'Osireion' of Seti I at Abydos, which is nine feet wide, there are no fewer than five rings of special arch-bricks.

In these composite arches, the spring of each successive ring is generally

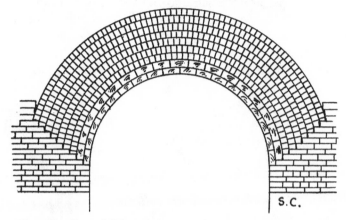

Fig. 216. Great brick archway, the inner rings, constructed of special arch-bricks, forming a centring for rings of ordinary bricks laid as in modern arches.

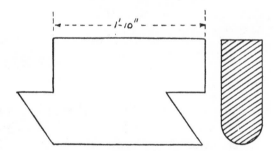

Fig. 217. Interlocking arch-brick from the offer- ing chamber of the tomb of Sabef. Old Kingdom. Gîza.

slightly higher than that of the one below it (Fig. 216), giving a structure of very great strength.

Arches constructed without centring are known from the Old King-dom. A very curious form of interlocking arch-brick belonging to this period was found in the offering-chamber of a man named Sabef at Gîza (Fig. 217).

Two forms of roofing gave rise to false arches in stone; the first is the pointed roof formed by pairs of large slabs leaning against each other, gable-wise. The uppermost relieving chamber of the King's Chamber in the Great Pyramid and the burial chamber of the Pyramid of Unas are roofed in this manner and the same form of roofing is frequently seen in the burial chambers of the Middle Kingdom,[1] during which period the under surface of the slabs was sometimes cut to give the effect of an arch. Considerable ingenuity is often shown in the manner in which the top of the slabs is treated in order to prevent them from slipping. Good examples of

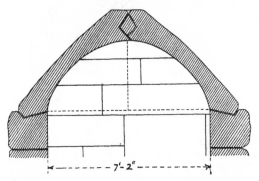

Fig. 218. False arch, cut from two granite blocks, in a Middle Kingdom mastaba at Dahshûr. (From De Morgan, *Fouilles à Dahchour, Mars-juin,* 1894, p. 55.)

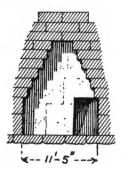

Fig. 219. Corbelled roof in the chamber of a XIIth dynasty mastaba at Dahshûr. (From De Morgan, *op. cit.,* p. 9.)

this form of false arch can be seen in the chamber of the XIth dynasty temple of El-Deir el-Bahari,[2] in the Pyramid of El-Lahûn,[3] and in some of the mastabas at Dahshûr (Fig. 218). In the New Kingdom it is no longer found.

The second form of roofing from which false arches were formed was the corbelled roof.[4] Corbelling consists of laying each course, after the wall of the chamber or passage has attained the required height, somewhat further forward than the one below, the process continuing until the opposite courses meet, or until they are close enough to be spanned by a single roof-

[1] Gautier-Jéquier, *Fouilles de Licht,* p. 71.
[2] Naville, *The XIth Dynasty Temple of Deir el-Bahari,* ii, Pl. XXII.
[3] Petrie, *Lahun,* ii, Pl. XXV.

[4] Corbelled roofs in brick, covering small graves, are known from very early times. See Reisner, *The Early Dynastic Cemeteries at Naga ed Deir,* i, p. 42.

block. The earliest corbelled roofs of any size are those of the chamber in the pyramid of Sneferu at Meydûm, and in the Grand Gallery of the Great Pyramid (Fig. 228). In the Middle Kingdom, corbelling is frequently seen in the roofing of mastabas (Fig. 219). In the New Kingdom, the principle was freely used for the construction of false arches, though the number of corbelled courses is far less than in the Old Kingdom

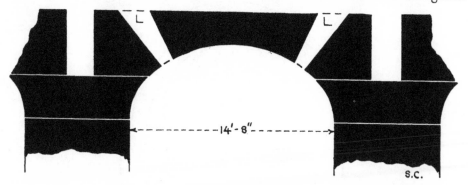

Fig. 220. False arch in the sanctuaries of the temple of Seti I at Abydos. *LL*, light-openings through the roof-slab.

Fig. 221. Corbelled arch in the central sanctuary of the XVIIIth dynasty temple at El-Deir el-Bahari.

examples. Such arches can be seen in the seven parallel sanctuaries in the temple of Seti I at Abydos (Fig. 220), in the central sanctuary of the XVIIIth dynasty temple at El-Deir el-Bahari (Fig. 221), in the temple of Ramesses III at Medînet Habu and in many other temples of the period.

The ancient Egyptians, like their modern descendants, could construct small brick cupolas or domes without the aid of centring. They were used to cover small chambers and kilns[1] and occasionally for roofing tomb-

[1] Mariette, *Abydos*, ii, Pls. LXVI & LXVII.

shafts to prevent them from becoming filled with sand.[1] They were constructed in rings, the bricks composing each ring being slightly corbelled
over those below, and at the same time tilted forward with the aid of sherds

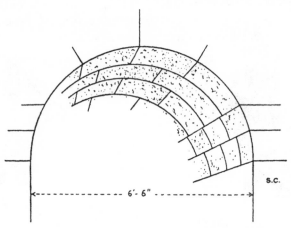

Fig. 222.   Arch in the eastern sanctuary of the XXVth-
dynasty shrines at Medînet Habu.

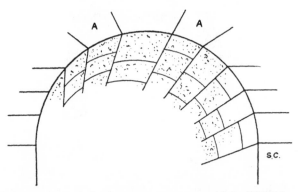

Fig. 223.   Arch in the western sanctuary of the XXVth-
dynasty shrines at Medînet Habu.

and limestone chips. At the time of writing (May 1927) a small chamber
with a domed roof, dating to the Old Kingdom, has been discovered by
the Vienna Academy of Sciences (Dr. Junker), but exact details are not
yet available for publication.

The arch having its crown held by a keystone does not make its appear-

---

[1] ENGELBACH, *Riqqeh and Memphis*, VI, p. 7.

ance until Saïte times. Even then it cannot be considered a true arch, since it is still largely dependent on corbelling. At Medînet Habu, there are three small shrines of the XXVth dynasty with arched roofs to their sanctuaries, both of which show special peculiarities. These shrines are in a perfect state of preservation, the part above the arches being covered by a roof, hence these interesting examples cannot be studied as fully as could be wished. The most that can be done is to determine the angle of the joints by inserting the blade of a knife into them where they gap sufficiently to permit it. In the first arch (Fig. 222), the four courses above the spring have been corbelled, and the joint between the uppermost of them and the next blocks is sufficiently flat to have permitted the next course (*A*) to be

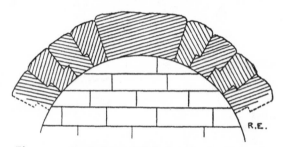

Fig. 224. Joggled arch of Ptolemaic date from the tombs at Kôm Abu Billo.

retained in position while the keystone was inserted. In the other two arches (Fig. 223), three courses were corbelled, the upper of these courses having a much greater depth than the others. Here, it cannot be determined whether the pair of keystones in every alternate ring was placed in position at the same moment or whether each was supported on the course below by joggling, cramps, or other means.[1]

The joggled arch in stone, where each voussoir except the keystone hangs on to that below owing to the peculiar shaping of the bedding joints (Fig. 224), is not known before Ptolemaic times, when it is found, among other places, in the tombs at Kôm Abu Billo in the Delta. In these tombs, which are of limestone, the arch is rather ovoid than semicircular, with a span of some seven feet. The voussoirs are rough on the outside of the tomb, and the tool-marks on the curved roof show that the surfaces inside were dressed after the blocks had been laid.

[1] The use of dovetail-cramps to keep two voussoirs in position during the construction is known in a late arch in the innermost hypostyle hall of Luxor Temple.

Even as late as the reign of Diocletian, the Egyptian method of constructing a stone arch without centring is maintained, and in the north-east
corner of Philae it is found applied to a miniature triumphal arch of classic

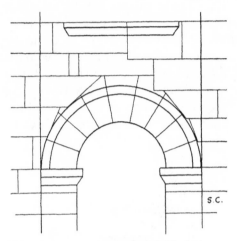

Fig. 225. Triumphal arch of Diocletian at
Philae.

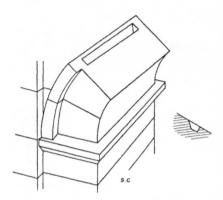

Fig. 226. Voussoir of one of the destroyed arches in the triumphal archway of Diocletian at Philae.

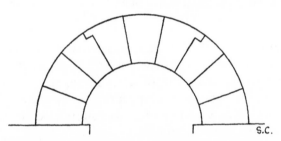

Fig. 227. Diagram showing method of laying the
fourth voussoirs in the triumphal arch of Diocletian
at Philae.

type—an archway just over four feet wide, side flanked by smaller ones. It
is clear from an inspection of the joints in the masonry of these arches (Fig.
225) that the blocks have been laid with the outer faces rough, and have
been dressed after they were in position. One of the arches is partly destroyed, and its voussoirs can be studied. In the upper joint of the first
voussoir a groove has been cut (Fig. 226) into which a corresponding projection on the lower face of the second voussoir is made to fit. The second

stone is thus prevented from slipping over the first. The third stone is attached to the second by similar means. The manner of attaching the fourth voussoirs can be seen from the inside of the perfect arch (Fig. 227). They are cut with a small projection on the outer edge of their joints by which they hang on a corresponding ledge in the joint of the third pair.

The Egyptians understood the necessity of relieving the pressure of super-incumbent masonry on the roofs of their buildings. In the Great Pyramid, above the massive granite roof-blocks of the King's Chamber, there are

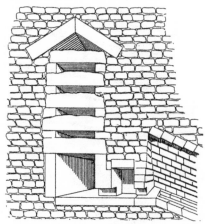

Fig. 228. Section of relieving chambers above the King's Chamber in the Great Pyramid. (After PERROT & CHIPIEZ, *Histoire de l'Art dans l'Antiquité*, p. 227.)

four low chambers similarly roofed (Fig. 228), and above these again is still another chamber roofed by pairs of huge blocks of limestone leaning against each other. It is a debatable point whether such a number of reliev-ing chambers is necessary; it certainly would not be needed in an inter-bonded mass of masonry, but the cores of pyramids are of roughly dressed blocks without any bonding, and, further, the straight joints of the accre-tion faces, if they exist in this pyramid (p. 123), would tend to produce a great vertical pressure on any roof chamber below, since they would pre-vent the weight of the upper courses of masonry from being evenly spread over the whole area of the course at which the chamber was built. In the Vth dynasty pyramids at Abusîr, as many as three pairs of blocks meeting at a point one above the other were considered necessary to protect the chamber below (Fig. 135).

In the brick pyramid of Hawâra, of the XIIth dynasty, the precautions taken to relieve the pressure of the brickwork on the masonry are even more elaborate. The walls and bottom of the chamber are, we are informed (p. 23, foot-note), of one piece of quartzite, and the former are heightened by one course of blocks. The chamber is roofed by three great horizontal beams of quartzite resting on the sides. Above this it is again roofed by longitudinal beams resting on blocks along the edges of the lower roof.

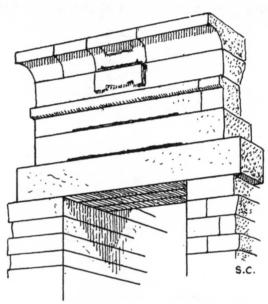

Fig. 229. Relieving device in the gateway of
Nectanebos II at Karnak.

Above this again there is a pointed roof composed of great pairs of lime-stone blocks leaning against each other, which must weigh nearly fifty tons each. All the blocks were well dressed at the joints and mortared together. The pointed roof rests on the masonry filling built up round the sepulchre. To make assurance doubly sure, an arch of brick, six courses deep, was constructed over the pointed roof. The courses of the arch were laid alternately in stretchers and headers, and the space between the arch and the pointed roof was composed of bricks laid in mud mortar. Above the arch the bricks were laid in sand. Fuller details of the relieving arch could not have been obtained without very extensive excavation in the mass of the pyramid.

As the practice of erecting vast piles of masonry above royal burials fell into disuse in the New Kingdom, the relieving arch, necessary in such structures, also disappeared. In none of the temples were they necessary. In the great gateway of Nectanebos II in the eastern temenos wall at Karnak, however, an effective means of preventing the breaking of the great lintel-blocks was taken, which may be considered as a relieving arch. The lower joints of the blocks which were placed above the lintel-blocks were slightly recessed so that they should only bear on the blocks below them at the parts which rested on the jambs (Fig. 229). Thus, in the middle, each beam had only to sustain its own weight. This practice was not continued by the Ptolemies in their great gateways at Karnak (Figs. 192 & 193). The lintel of one of these has failed.

It may well be asked why the Egyptians, understanding the principle of centring, did not apply it to stone arches, especially since it was the custom, in constructing a building of any size, to surround it with embankments and fill the interior with earth. In such cases it would, to all appearance, be a simple matter to make a centring of earth and to build the arch upon it. The invariable Egyptian custom of laying a block with its face in the rough would have made centring almost impracticable for a stone arch, since the rough face on the soffit of the arch would have rested on the centring and prevented the voussoir from being properly laid; further, a corbelled arch was quite as simple, if not simpler, to construct.

It should be noted that the Egyptian arches which have a keystone, like those of Medînet Habu and Philae, are all in comparatively small buildings built of small blocks. It is very likely that, if the blocks were small enough to be handled by three or four men, the cumbersome constructional embankments were not used at all.

XVIII

FACING, SCULPTURING, AND PAINTING THE MASONRY

THE earliest temples known to have been sculptured and painted are those at Abusîr,[1] where the work, in low relief, is of exceptionally fine quality, as regards both proportion and colour. The IVth dynasty temples of the pyramid of Sneferu at Meydûm and those of Gîza are perfectly plain, though the tomb-chapels of the same period in both places are profusely adorned with scenes and inscriptions.[2]

Very little is known of the temples of the Middle Kingdom, but, by the New Kingdom, a temple was not considered complete until all the surfaces of the masonry had been sculptured and painted. Some of the temples, however, did not reach this stage.

An examination of the temples of the New Kingdom gives the impression that there can have been no collaboration between the architects and the artists, and joints sometimes occur in the most unsuitable places on the scenes (Fig. 150). It must be remembered, however, that when the temple was complete the joints in the masonry did not show; everything was covered with gesso, on which the painting was done. For many years all looked well, but there must have been plenty of examples in the temples of past reigns of what would be the ultimate result of this careless manner of procedure, when the gesso had begun to flake off. This lack of foresight is one of the most remarkable traits in the character of the ancient Egyptians.

The practice of the Egyptians of laying their blocks with the minimum number of dressed sides had the result of leaving the face of the masonry of a building, when the blocks had all been laid, quite rough.

Fortunately, unfinished buildings of nearly all periods have been preserved, both in megalithic and small-block masonry. Among the more important of these are the granite casing of the Third Pyramid at Gîza (Figs. 99 & 100), the wall of the court of Amenophis III in Luxor Temple (Fig. 233), the quartzite walls of the 'Osireion' of Seti I at Abydos (Fig. 81), the First Pylon at Karnak of the XXVth dynasty (Fig. 87),

---

[1] BORCHARDT, *Das Grabdenkmal des Königs Sahurĕ'*.
[2] Some of the chambers recently discovered in the royal mastaba of the IIIrd dynasty at Saqqâra, within the boundary wall of the Step Pyramid (Fig. 91, p. 96), are decorated with fine reliefs of King Zoser. It is not yet certain (May 1927) what the purpose of this mastaba may have been.

and the unfinished repairs undertaken in the small shrines of Seti II. When it is understood that the face of the masonry was the last to be

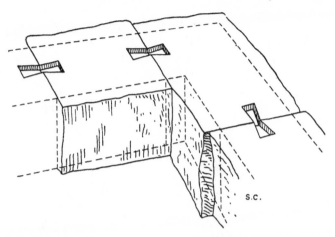

Fig. 230. Diagram showing that, in Egyptian buildings, the rising joint between the blocks does not necessarily occur at the interior corners of a wall, since the front face of the masonry was dressed after the blocks were laid. 'Temple of the Sphinx'; Gîza.

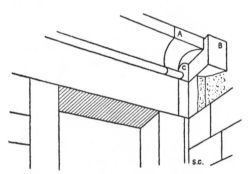

Fig. 231. Doorway in the temple of Seti I at Abydos. The parts *A*, *B*, and *C* are all contained in one block of stone.

dressed, the curious position in which the joints of the masonry are found with regard to the architectural details is easily comprehensible. In the valley temple of the Second Pyramid at Gîza, for instance, the rising joints of the blocks are rarely at the internal corners of the building (Fig. 230), and, to take an extreme example from the temple of Seti I at Abydos (Fig. 231), the end of the cornice of a doorway, part of the moulding and the corner of the flanking wall are all cut from one block.

The tools used in dressing the faces of the masonry were, for the hard stones, the dolerite pounders and mauls (Fig. 266), and a blunt-pointed tool, perhaps akin to the mason's pick, and for the soft stones, the chisel struck by the mallet (Fig. 263), and, in later masonry, some form of adze. For the harder limestones the pounders and the pick were used as well as chisels. Both the hard and the soft rocks were at times drilled and sawn (p. 203). No scene has been preserved where the dressing of

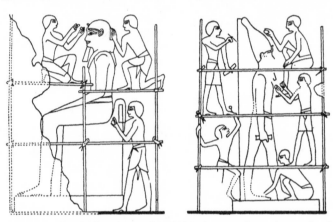

Fig. 232. Scaffolding in the XVIIIth dynasty, from the tomb of Rakhmirē' at Thebes. (After NEWBERRY, *Rekhmara*, Pl. XX.)

the face of a wall is depicted, but there is a certain amount of evidence to be gathered from the monuments themselves and by the elimination of impossibilities. In the pyramids, we are led to suppose that the surfaces were dressed, during the removal of the constructional embankments, to the plane of facing-surfaces left while the courses were being laid (p. 124).

In the two examples of hard-rock masonry where the blocks have never been faced—the casing of the Third Pyramid and the walls of the 'Osireion' —no 'facing-surfaces' are found. In the former case it was doubtless the intention of the masons to dress the surface of the granite to the plane of the finished limestone casing above (p. 128) and in the latter, since the whole surface could easily be scaffolded, the process could wait the convenience of the builders. As it happened, the dressing of the 'Osireion'

Fig. 233. Unfinished wall in the temple of Amenophis III at Luxor. (The back of this wall is shown in *Fig.* 127)

Fig. 234. The unfinished pylon (No. 1) at Karnak, from the north-east, showing remains of the constructional embankments

Fig. 235. Masonry of the unfinished pylon at Karnak

Fig. 236. Masonry of the unfinished pylon at Karnak

seems to have proved too laborious for either Seti I or his successors to complete.

The outer surface of the wall which bounded the court of Amenophis III at Luxor Temple on the east and south-east has never been faced (Fig. 233). At intervals of about forty feet on the east wall and at two points at the south-east corner, there are what appear to be vertical facing-surfaces running up the masonry. Though they have not been fine-dressed, and seem to be unfinished, they are at the batter to which the wall was to be finally dressed, judging from that of the corresponding wall on the west side. They may well have been cut into the masonry by measurement from a plumb-line as the first step to dressing the whole surface. It is not clear why this face of the court-wall was never completed, and the outer faces of the whole of this side of the temple of Amenophis III hardly sculptured at all, either by him or by his predecessors. In Egypt, such unfinished work is constantly encountered—especially in tombs—and is most difficult to explain satisfactorily.

It is probable that in all but the highest structures, such as pyramids or pylons, the dressing and sculpturing was carried out by means of scaffolding, with which the Egyptians were well acquainted (Fig. 232).

The procedure for reducing the whole surface of the masonry to the plane of the facing-surfaces was no doubt similar to that used for flattening a single block (p. 105).

The great unfinished pylon at Karnak furnishes several interesting hints on the manner in which the dressing was to have been carried out. There is little doubt that it was abandoned with the constructional embankments all round it, and that these have largely been removed in the course of ages for *sibâkh*, or manure to spread on the fields.

The blocks of which it is composed are drafted, but this drafting is very different from that seen in medieval and modern masonry. In the first place, the quarry surfaces still on the faces of the blocks are vertical, while the 'drafts' are very much deeper at the top of the blocks than at the bottom (Fig. 234). Further, the top drafts of a line of blocks were cut *after* they were laid, since the tool-marks on them frequently pass from one block to the others. In the places where the top of a line of blocks is exposed for any considerable distance, it can be seen that the line of the top draftings on them is perfectly straight as far as can be traced, which seems to indicate that the line was obtained by measurement from plumb-lines at each end of the towers, and marks the amount of stone to be removed in order to obtain the required batter of the pylon, which is about a rise of 7 on a horizontal distance of 1. In other words, it is almost certain that the top drafts of the blocks of each course are the 'facing-surfaces' to which the whole face would afterwards be dressed. The lower draft of the blocks seems to have been cut before laying, and the intention was most likely to lay a block so that its lower edge lay as nearly as possible on that of the block below, the drafting on its lower edge being to expose the facing-surface over which it lay. The 'drafts' between an upper and a lower block, however, by no means invariably coincide (Fig. 235, *A*), and here and there they have been re-cut in order to make them do so. Whether the drafts on the rising joints of the blocks were cut before laying is uncertain. They rarely coincide

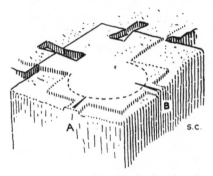

Fig. 237. One of the blocks forming the torus moulding in the unfinished pylon at Karnak, showing recessing of the block and masons' guide-lines.

with those of the neighbouring blocks, but on one side of the rising joint the draft has been cut to make a straight strip between the upper draft of its own block and that of the block below it, thus forming a vertical 'facing-surface' (Figs. 235 & 236, *B*). It thus appears that when the blocks forming the pylon had all been laid, the masonry behind the constructional embankments was covered with facing-surfaces to the plane of which the whole face of the pylon could easily be dressed while the embankments were being removed, which is what has already been postulated in the case of pyramids (p. 124). The dressing seems to have actually taken place for the gateway, and it is likely that the embankments were removed from this part of the pylon as soon as possible in order to clear the entrance into the temple. Why the remainder of the pylon was never faced will probably never be known for certain. In this pylon the torus moulding is constructed with the same precautions for the final dressing (Figs. 237 & 238). The

Fig. 238. Surplus stone left at the corner of the unfinished pylon at Karnak for the torus moulding

Fig. 239. Unfinished repairs to the small temple of Seti II at Karnak

draft lines along the top of the courses are continued on to the block at the corner which is to form part of the moulding, where a square projection was allowed for, which could afterwards be cut down to form the required circle. Guide-lines are left to assist in the setting of the block of the next course.

Two other remarkable features can be seen in the masonry of this pylon; the first is the presence of recesses cut down here and there into the top of the rough, projecting quarry-face on the blocks. As many as three such recesses are sometimes found in a single stone. Their position, and the fact that their presence is only occasional, shows that they were not used in connexion with any lifting appliance such as tongs. We suggest that they were

Fig. 240. Workmen dressing the surface of a stone offering-table with pounding-balls; from the XVIIIth-dynasty tomb of Rakhmirē‘. (After NEWBERRY, *Rekhmara*, Pl. XX.)

Fig. 241. Men polishing and working with a stone chisel on a sphinx, from the tomb of Rakhmirē‘ at Thebes; XVIIIth dynasty. (After NEWBERRY, *Rekhmara*, Pl. XX.)

merely for holding the workmen's tools, since similar cuttings in the quarries at Tura and Ma‘sara can often be seen. The second feature consists of large lumps of white mortar adhering at very irregular intervals to the bedding joints of the blocks (Figs. 235 & 236, *C*). If one of these is removed, the flatness of the drafting below it seems to be of unusual excellence. We can suggest no explanation of these lumps of mortar. This pylon is well worthy of a very detailed study, since a great deal of new information might well be gathered from it.

The western part of the small temple of Seti II at Karnak seems to have been repaired at the same time that the unfinished pylon was commenced, namely, in the XXVth dynasty, and, with the rise in the masonry of the pylon, it became buried within the constructional embankments (Fig. 239). It was, no doubt, intended to dress the faces of the repaired part when the constructional embankments were removed, but this, for some unknown reason, never took place. The repairs are exceedingly rough; the stones are neither well set nor well arranged. Many of the stones are 'drafted' like those of the pylon, and the bedding joints are mostly horizontal.

When the whole surface of a wall or pylon had been flattened, it may have been given a rubbing with polishing stones. Large pieces of hard sandstone with a flat surface on them are often found, which may have been used for this purpose. For the hard rocks, some form of polishing powder must have been used in conjunction with the polishing stones. The tomb of Rakhmirē' at Thebes contains two scenes depicting the dressing and

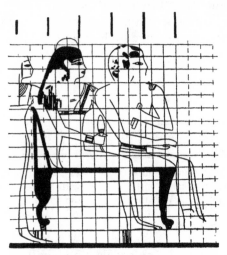

Fig. 242. Proportion squares in the tomb of Wah (No. 22) at Thebes. From MACKAY, *Journal of Egyptian Archaeology*, iv, pp. 74–85. (Block kindly lent by the Editor.)

polishing of what are probably monuments of hard stone; in the first (Fig. 240), three men are using the dolerite pounding-balls for dressing the surface of an offering table, while the foreman, stick in hand, seems to be inquiring why the work is not proceeding more quickly. In the second scene (Fig. 241) a man is pounding out as much detail as he can into the uraeus on the head of the Sphinx, using a piece of stone—probably also of dolerite—having a chisel edge, in order to get as far as possible into the angles. On the back of the Sphinx a man is rubbing with a polishing stone while a colleague is standing by with a bowl and a brush, probably containing the polishing powder which he is ready to smear on the stone. Polishing stones are known in granite, in basalt, in quartzite, and in sandstone (Fig. 266). When its surfaces had been dressed, the building was complete as far as the architects were concerned, and the artists began their work. In cases where it was required to cover the wall with scenes in relief, the surface was first covered with squares, generally in red, drawn by plucking a stretched cord dipped in ochre (Fig. 266) which was held against the surface. These squares not only gave the vertical and horizontal and enabled the inscriptions to be properly spaced and maintained straight, but they also permitted the figures to be drawn to the proper canon of proportion, which was subject to fairly fixed rules in the different dynasties[1] (Figs. 242 & 243).

The draughtsman then sketched the scenes and inscriptions and either he or another draughtsman gave a clear outline to the whole scene. Correc-

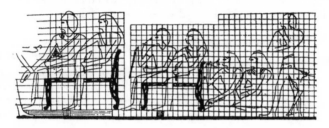

Fig. 243. Proportion squares in the tomb of Tati (No. 154) at Thebes. From MACKAY, *Journal of Egyptian Archaeology,* iv, pp. 74–85. (Block kindly lent by the editor.)

tions, presumably by a superior hand, are often seen at this and other stages of the work (Fig. 244). The sculptor followed the draughtsman, and cut away the ground to a depth of about one-eighth of an inch, so that the figure stood out, square edged, on the ground or field. The man who reduced the field appears to have been followed by a superior sculptor, who rounded off the edges of the figures and put in some modelling on the limbs and faces of the figures, and, in the best work, on the hieroglyphs. A coat of gesso—which is a mixture of whiting and glue—was then put over the whole work, probably with a brush; this was followed by a coating of size, and the scene was ready for the painter. In painting the small details of the scenes and inscriptions, the artist was not always very careful in following the details indicated by the sculptor, but in general the work done by him is admirably even and delicate.

We are indebted to Mr. A. Lucas, late Director of the Chemical Depart-

---

[1] MACKAY, *Journal of Egyptian Archaeology*, iv, pp. 74–85.

ment and now working for the Antiquities Department, for the following notes on the composition of the ancient pigments:

'Most of the Egyptian pigments were naturally-occurring mineral substances, simply powdered.

'The white was generally carbonate of lime, but sometimes sulphate of lime.

'The black was carbon, being sometimes soot and sometimes a coarse material, probably powdered charcoal.

'The grey was a mixture of black and white.

'The red was red ochre, either natural or made by calcining yellow ochre. In Roman times, however, red lead was also employed as well as pink made from madder.

'The browns were all natural ochres.

'The yellow was of two kinds, one being natural yellow ochre and the other, which, however, was not used until about the XVIIIth dynasty, was sulphide of arsenic (orpiment), and as this latter does not occur in Egypt, the supply must have been imported.

'The principal blue was an artificial frit, consisting of a crystalline copper-lime-silicate made from malachite, limestone, and powdered quartz pebbles, possibly with the aid of natron—though this latter was not necessary. This is known as early as the XIth dynasty. Another and earlier blue was powdered azurite, a naturally occurring basic carbonate of copper, which was used before the artificial frit was discovered. Still another blue, the occasional use of which has been reported, was a cobalt compound.

'The green used was of two kinds; that employed at first being powdered malachite—a copper ore found in Egypt—and at a later date a green frit, analogous to the blue frit already mentioned.

'The medium with which the colours were put on was water and not oil, with size, gum, or white of egg. It has not yet been definitely established which of the three was used.

'The Egyptian painting was in reality a distemper.'

Imitative painting is known in Egypt at a very early date. The lower surface of the roof-slabs in the IIIrd-dynasty tomb of Hetephernebti was painted to represent logs. Dr. G. Reisner has kindly sent us the following note on his work at Gîza:

'The painting of chapel walls and roofs to imitate red granite is quite common. In one of the tombs recently discovered by us, the whole offering room was cut in the solid rock (nummulitic limestone) and painted on the roof and walls to imitate red granite with the exception of the seams on the west and north walls, which were sized with plaster and modelled to imitate fine white limestone. I know of three other tombs in our work at Gîza, and a stele, on which the imitative colouring of granite is well preserved. All were of Dyn. V or Dyn. VI; none were of Dyn. IV. The imitation of wood graining also occurs, but only on the "false door", and more often on plastered crude brick than on stone. The imitation of wooden roofing (logs) occurs

Fig. 244. Correction in the pose of a king after the scene had been sculptured. Temple of Amenophis III at Luxor

Fig. 245. Eye and eyebrow of a small obsidian head of King Tuthmosis III, from Karnak, showing method of cutting the stone by a line of drill-holes. Full size

in both stone and crude-bricks. The two cases on crude brick were both in "leaning-course" vaults. The wood was painted red (no graining) in all cases.'

In the XIth-dynasty temple at El-Deir el-Bahari, some very delicate and interesting specimens of imitative painting were revealed during its clearance.[1] Here the graining of the wood is painted on highly elaborated shrines and occasionally on parts of the temple structure.

Of the personalities of the artists to whom the sculpturing and painting of the monuments is due, we know comparatively little. The 'outline draughtsmen', sculptors—or 'wielders of the chisel'—and painters seem to have been of the middle classes, to judge from those whose tombs are known in the Theban Necropolis. None of the highest titles is associated with those borne by the artists. At Deir el-Medîna there is a small collection of graves of sculptors, but the bad quality of the rock in which they were permitted to make their tombs prevented them from giving good samples of their art. In these tombs there seems to be no case where a sculptor has also the titles of painter or outline-draughtsman. Each man seems to have had his special job, and they may well have had some sort of guild, though there is no proof of it in dynastic times.

Many visitors to the monuments express surprise that the painting could have been carried out in the darkness of the tombs and in the dim light of the temples. The Egyptian lamp was of the simplest type, merely a wick floating in oil. It is not infrequently represented in the scenes in the tombs, where it usually takes the form of an open receptacle mounted on a tall foot which, in the smaller examples, can be grasped in the hand. In the pictures, there arise from the receptacle what we may assume to be wicks or flames, always curved over at the top as if blown by a current of air.[2] Stand-lamps in limestone have been found in the pyramid of El-Lahûn,[3] and representations of them in stone in the 'Labyrinth' at Hawâra.[4] In Egyptian houses, small dishes were also used as lamps. They usually have their rims pinched into a spout to enable the oil to be poured out if necessary. Dishes used as lamps can be seen in the Cairo Museum. The absence of smoke-blackening in the tombs of the Kings is also not difficult of explanation. If olive-oil is used, there is very little smoke, and a suitable covering over the lamp, for which various methods readily suggest themselves, would very easily prevent carbon being deposited on the ceiling.

[1] NAVILLE, *The XIth Dynasty Temple at Deir el Bahari*, ii, Pls. XII, XIII, XV, XVI, XVIII, and XIX.
[2] TYLOR and GRIFFITH, *Wall Drawings and Monuments of El-Kab* (Tomb of Paheri), Pl. XIII.
[3] BRUNTON, *Lahun*, i, *The Treasure*, Pl. XX.
[4] PETRIE, *The Labyrinth, Hawara and Mazghuneh*, Pl. XXVIII.

Though the cutting of hieroglyphs in the soft rocks presents few difficulties to the sculptor, that on, for instance, a granite doorway is a very different matter. Pounding with balls of dolerite, or even with pointed stones of the same material, was of little avail where sharp corners and narrow channels were required, as, for example, in incising hieroglyphs. Though the final stage of the sculpturing of the hard rocks, apart from the polishing, was often carried out entirely by means of a pointed implement, such as a mason's pick, in cases where the width of the channel would not permit the use of such a tool, or where the material to be cut was extremely brittle, a line of little drill-holes, close together, was sunk, and the intermediate portions chipped away (Fig. 245). The ancient Egyptians could drill the hardest rocks, but much is yet dark regarding the actual technique. The interest of ancient drilling was first brought to the notice of the public by Sir Flinders (then Mr.) Petrie as far back as 1883, at a lecture to the Anthropological Institute,[1] since when no advance of any moment seems to have been made in the understanding of the processes used. In the lecture referred to, he cites some examples of drilling in the hard rocks, three of which are especially worthy of consideration. The first is a granite drill-core, which is grooved round and round by a graving-point. The grooves form a spiral, and in one part a single groove may be traced around the core for a length of five rotations, equal to three feet. He remarks that even at the ends of this groove there is no sensible difference in its character, as there would be, for instance, if the cutting-point had begun to fail. At the end of the spiral the grooves become confused, owing merely to the irregular action of the tool. The second specimen is part of a drill-hole in diorite. The hole has been $4\frac{1}{2}$ inches in diameter, or 14 inches in circumference. As seventeen equidistant grooves appear to be due to successive rotations of the same cutting-point, a single cut is thus 20 feet in length. The third specimen—a piece of diorite—shows a series of grooves, each ploughed out to a depth of $\frac{1}{100}$ inch at a single cut without any irregularity or 'starting' of the tool.

Sometimes very large tubular drills were used, even on the hard rocks. A good example can be seen in the fine diorite statue of King Kha'frē' in the Cairo Museum, where the drilling process, used for the removal of the material between the legs, has been carried down too deeply. From this it is clear that the diameter of the drill was nearly two inches.

A hard stone can be drilled by two methods; one is by a tool revolving at

[1] PETRIE, 'On the Mechanical Methods of the Ancient Egyptians' (*Journal of the Anthropological Institute*, August 1883).

a very high speed, and the other by great pressure on a slow-moving tool of great hardness or in conjunction with a cutting powder. As an example of the effect obtained by high speeds, one of the writers, at Messrs. Farmer and Brindley's marble works, saw a piece of biscuit-tin, formed into a cylinder with a ragged edge, fixed into a high-velocity drill, by means of which a hole was easily sunk in a piece of porphyry. The Egyptians had no high-velocity drills; the greatest speed obtainable with a small drill was by rotating it by means of a bow, exactly as the Egyptian carpenters do to-day.

Fig. 246. Man drilling out the interior of a stone vase. Vth dynasty; from Abusîr.

It is fairly certain that in ancient times a considerable weight was put upon the drill. In a scene from the Vth-dynasty temple of Sahurē' at Abusîr,[1] a man is depicted drilling out the interior of a stone vase (Fig. 246). The top of the drill-shaft is fitted with a curious eccentric handle and has heavy weights attached to it. He is rotating the drill both by the handle and by pushing the weights round with his hand.

Not only could the Egyptians drill the hardest rocks, but they could also saw them. The sawing of the hard rocks presents very similar problems to the drilling. In the basalt triads of King Menkewrē' of the IVth dynasty, one has not had the back smoothed, and on it saw marks can be clearly seen (Fig. 247). Some of the scorings made by the saw are nearly an eighth of an inch apart, though here it is incredible that they represent successive

[1] Now in the Cairo Museum.

strokes of the tool. The granite coffer of the Great Pyramid also exhibits traces of saw-cuts. On this Professor Petrie remarks that the saw must have been over eight feet in length, since the grooves run lengthways on the side of the coffer, which is 7 feet 6 inches long, and some length of stroke must be allowed for. In this, the grooves are very much closer than those in the triad of Menkewrē'. Saw-cuts are known as narrow as 0·03 inch, which almost certainly rules out the possible use of a blade with fixed jewels set in it.

To-day, granite is sawn by means of small chilled shot, the movement being imparted by a blade, without teeth, a quarter of an inch thick. It is also possible to cut granite, porphyry, and other hard rocks with sand instead of shot, but this is a very slow process. One of the writers has cut granite, using fragments of emery about $\frac{1}{10}$-inch 'diameter' in conjunction with a steel saw with its teeth almost ground away and its temper destroyed. The emery was used wet, and as much weight as possible put on to the saw, which was raised every few strokes to enable fresh emery to penetrate into the groove. Though the cut thus made showed longitudinal marks on the side of the groove, it must be admitted that they were not as deep nor as far apart as those in the ancient examples. Experiments with quartz instead of emery gave almost identical results.

It is quite likely that the ancient saws and drills were toothed, not for cutting the stone, but to move whatever cutting agent may have been used along the groove. By experiment alone, and the use of different cutting agents and different weights on the tool, can one hope to solve the problem of the ancient drilling and sawing. This could well be done by any one having access to a workshop and with sufficient leisure to devote to the subject.

In the soft rocks, when a channel was required to be made, it was sometimes drilled with specially made flint borers (Figs. 248 & 249) which were, no doubt, rotated by a shaft to which weights were attached (Fig. 246). Flint borers are known as early as the Old Kingdom.

It may not be out of place to consider what the appearance of the interior of a completed temple of the New Kingdom may have been. The general effect striven for was an increase of mystery as the successive halls—each smaller than the last—were traversed, and the climax of the effect was the small, dark chamber which contained the image of the venerated object. The interior of the temple did not, however, produce the impression of gloom which now strikes the visitor, since the ceiling, walls, and columns were all brightly coloured. The ceilings were usually painted blue and

Fig. 247. Back of a basalt triad of King Menkewrē', showing marks left by saw. Scale ½. Cairo Museum

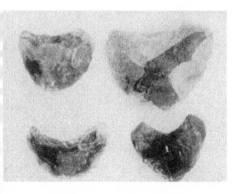

Fig. 248. Flint borers from Saqqâra; prob-bly Old Kingdom. (Photograph by C. M. Firth, Esq.)

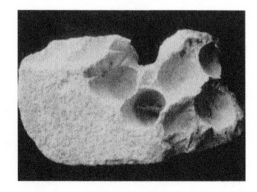

Fig. 249. Trial holes made with flint borers in a IIIrd dynasty stele from Saqqâra. (Photograph by C. M. Firth, Esq.)

studded with stars, and the walls and columns covered with inscriptions and scenes either in low or in sunk relief. Not only was colour freely used, but many conspicuous parts of the temple, such as doors, the lower parts of the door-jambs and even the abaci of some of the columns were, as the Egyptians put it, 'overlaid with gold' or 'overlaid with electrum'.[1] This expression seems to cover not only the nailing of thin gold plates on to wooden pegs driven into slots in the masonry, but also the more economical gilding by means of gold-leaf. The huge doors which divided one hall from its neighbour must have been a magnificent spectacle when 'overlaid' with gold, presenting a surface of some forty feet high by eight feet wide. What appears now as garish must have been softened down to an extremely pleasing effect when seen in the subdued light inside the temple.

Comparatively little is known of the appearance of the outside of the temples from the point of view of decoration, as the colour with which the sculptures were painted has mostly disappeared. There is no doubt, however, that, from the New Kingdom onward, many of the temples were as brightly painted outside as they were inside.

It is well to note that the Egyptians did not regard the scenes and the inscriptions on the temple-walls entirely as decoration. Certain objects and acts represented on the walls were thought to be created or to take place in the next world. The scenes on the walls seem sometimes to have served as records, as in the foundation scenes, as magical instruction to the deceased, as in the case of the Tombs of the Kings, and as actual offerings to the gods made by the king in whose honour the temple was built. That some had a purpose other than that of impressing the visitor with the glories of the king is very clear, since in such a case as the tomb of Seti I no person would ever enter it once the king was buried; yet the decoration is superb.

Though the Egyptians adorned their temples so sumptuously, they seem to have taken no pains, at any rate in late times, to keep them free from the incrustations of private mud-brick houses. This indifference—or perhaps toleration—can be seen in the mosques to-day, where sheds and shops are built against them freely. In the case of many of the later temples, houses were built against the mud-brick temenos walls, and in the course of centuries the level of the town reached almost, if not quite, to the top of the wall, the temple thus standing in a vast hollow in the middle of the town. The temples of Edfu, Esna, and Dendera are striking examples of this.

[1] Electrum was a silver-gold alloy. Both silver and copper were also used for 'overlaying'.

Herodotus, who visited Egypt about 460 B.C., remarks, on the temple of Bubastis (Zagazig):[1]

'The temple stands in the middle of the city and is visible on all sides as one walks around it, for, as the city has been raised by embankment[2] while the temple has been left as it was originally built, you look down on it wherever you are.'

It seems that the Egyptians at this time cared only for the effects of height and majesty in a temple to be felt when the spectator was inside the brick temenos wall.

[1] Book II, Chapter 138.
[2] Meaning that the houses have been built on the ruins of former houses, so that the level of the whole town has risen. The huge 'tells' or mounds in Egypt, some of which rise to a height of 60 feet or more, have been formed in this manner.

XIX

BRICKWORK[1]

IN the comparatively rainless climate of Egypt, the alluvial mud of the Nile, when made into bricks, forms a most convenient building material. Further, the mud itself can be used as the mortar. The bricks are not burnt, but merely dried in the fierce rays of the sun.[2]

The technique of brickmaking and bricklaying in Egypt has never changed from the earliest times until to-day, so that a study of modern methods gives great help in understanding those of ancient times. There is, however, one important side of ancient brickwork for the understanding of which modern methods are of little value, namely, the technique of the giant brick walls, which sometimes attain a height of forty-five feet and a thickness of anything up to eighty.[3] Such walls were constantly constructed by the Egyptians. It might well be imagined that to make a great brick wall would be more difficult than to build an ordinary house-wall, but this is not the case by any means. Many factors arise, to which a house-wall is not subject, which, curiously enough, tend to affect the solidity of a great mass of brickwork. The modern Egyptian does not build giant walls; there is no reason why he should. The small brick wall is always subject to damage by rain and by cracking, but it is astounding how much rain the mud houses in some of the Delta villages endure, though there is, as might well be expected, a limit, which was reached in the heavy downpours of the winter of 1923. Cracking can be minimized by using well-matured bricks and by certain other elementary precautions. On the great walls, rain had very little effect; wind-erosion played a far greater part in reducing them to the faceless condition in which most of them are seen to-day.

It is fortunate that good examples of ancient brickwork of practically every period of Egyptian history have been preserved, in spite of the gradual

---

[1] Brick arches are described in Chapter XVII.

[2] Mud-bricks are not always the most convenient building material in Egypt. In certain parts of Upper Egypt, above Esna, there is not sufficient good alluvium for making crude bricks in any considerable quantity, and the soil is often too sandy for the bricks to have any cohesive quality. As a compensation for this, the bad limestone gives place to a very good, soft, and durable sandstone, which has the advantage of breaking up into more or less rectangular forms. Stuck together with the alluvium as a mortar, all necessary materials for house-building are at hand.

[3] The wall of Tanis was of this thickness, and was all the work of King Ramesses II.

rise in the level of the alluvium, which has engulfed so many of the ancient sites. The limits of ancient Thebes with its hundred gates, as seen by the classical writers, can hardly now be traced, and Memphis, Heliopolis, Mendes, Buto, and a host of other towns, important in classical times, are now but heaps of brick débris, the top layers being of comparatively recent date. As a house fell, the unbroken bricks were salved; the remainder were broken up and a level floor made of them, and a new house was constructed which was consequently just a little higher, in level, than its predecessor. The result of this process, which went on for thousands of years, was that the towns in course of time stood on a hill, or *tell*, as the modern Egyptians call it. In some ancient sites the process is still going on; in others the inhabitants of the town have migrated to a more convenient spot, leaving the *tell* standing grim and deserted. Such a *tell* is a very valuable piece of property, since the ancient débris is used by the natives for manure, or *sibâkh*, to spread on the fields. The need of *sibâkh* has resulted in terrible havoc among the ancient *tells*. Bubastis and Athribis—to mention but a couple—have now been reduced nearly to cultivation level. Only a few, such as Buto and Tanis, have escaped extensive damage owing to their isolation, and they will no doubt provide a rich harvest to any excavators of the future who have sufficient funds—and patience—to bear some fruitless years of preliminary work while the upper layers of débris are being removed.

The method of making or 'striking' bricks used to-day by the Egyptian is worth description.

The brickmaker (Fig. 250), called *tawwâb* in Arabic, searches for a deposit of Nile mud of a suitable consistency for his purpose,[1] and clears as large and flat a space as possible. His assistants dig up the mud and put it into a smallish hole in the ground (*ma'gana* or *makhmara*), where water is added to it until it has the consistency of a very thick paste. The mixing is done with the aid of a cultivator's hoe (*fâs* or *ţûria*), the feet assisting in the operation. If chaff[2] (*tibn*) is available, it is mixed in varying quantities

[1] A very close examination of the soil is necessary before the great differences in their value for brick-making can be appreciated. On the same island one may observe that at one end of it there may be sand and at the other alluvium. Successive sections of the river bank show the same abrupt variations. The brickmaker has consequently learnt to make his own mixtures of mud and sand. With a good sand and alluvium he can make tolerably hard bricks, and dispense with the chaff which, in the southern

parts of Egypt, is expensive. Burning material, such as maize-stalks (*bûş*), is also costly, hence burnt bricks are not used to-day in the country unless it be by the rich 'notable' who wishes to make a good show. It may also be remarked that camel and donkey manure, the common fuel of the countryman, is useless for burning bricks.

[2] Obtained from threshing-floors after the oxen have trodden the corn. Manure is occasionally added, giving an exceptionally tough brick.

Fig. 250. Modern brickmaker and his assistant; Luxor. (Photograph by Gaddis and Seif)

with the mud paste; if there is no *tibn* the bricks are made without it, but sand is often added with good effect. Having thoroughly mixed up the paste, an assistant takes a round or oval mat (*bursh*) made of strips of palm leaf (*khûs*), having handles on either side, and, having dusted it over with fine dry mud to prevent sticking, he puts as much of the paste on it as he can carry and leaves it beside the brickmaker. The brickmaker squats down, holding an oblong wooden mould fitted with a handle—of which examples are known in the XIIth and the XVIIIth dynasties (Fig. 263, p. 224)—the mould being of the size of the bricks he wishes to 'strike' (*darab*). Having filled the mould with the mud paste, the brickmaker scrapes off the surplus and lifts off the mould, leaving a sticky mud brick, just sufficiently hard to retain its form. He continues 'striking' a series of

Fig. 251. Captives making bricks for the storehouses of the temple of Amūn; XVIIIth dynasty. (After Newberry, *Rekhmara*, Pl. XXI.)

such bricks, one alongside the other, until all his available space is filled. The bricks must then be left to dry until they are hard enough to be stacked and a new series made. In ancient Egypt the method was identical with that used to-day, the only difference being that in the old scenes (Fig. 251) the mud is carried in a pot instead of on a mat. Nowadays, in addition to the mat, almost any receptacle is used, particularly petrol and paraffin tins.

Brick walls are built without scaffolding; the builders walk on the top of the piece they have built. Since they are barefooted their weight helps to solidify the work.

In ancient times two kinds of bricks were used, the house-bricks, which measure about 9 × 4½ × 3 inches, and the bricks used for government work, such as walls of towns, temples, and fortresses. The latter were of much greater size than the former, the average dimensions taken at random being as follows:

Nubian Fortresses (Middle Kingdom)   12 × 6 × 3 inches

El-Kâb (New Kingdom)	.	.	$15 \times 7\frac{1}{2} \times 6\frac{3}{4}$ inches
Karnak (New Kingdom)	.	.	$\begin{cases} 16 \times 8 \times 6 \quad \text{,,} \\ 14 \times 7 \times 5 \quad \text{,,} \end{cases}$
Armant	.		$\begin{cases} 14 \times 7 \times 4\frac{1}{2} \quad \text{,,} \\ 12 \times 6 \times 3\frac{3}{4} \quad \text{,,} \end{cases}$

In the construction of a great brick wall, precautions had to be taken to ensure that the interior of the mass dried as quickly as possible, and that there should be no cracking or shrinkage during the drying. In walls built on the alluvium, the uneven settling of the soil after the Nile flood had also to be taken into very serious consideration.

The drying was helped by leaving air-passages running transversely through the wall, which connected with similar passages running longitudinally. They were usually two bricks high and half a brick wide. In the great wall at Karnak the air-passages occur at every thirteen courses, or just over six feet up the wall, and are 2 feet 9 inches apart along the course.

Another precaution to ensure quick drying and to assist in preventing longitudinal cracking was to lay mats of reed or *halfa* grass, or loose rushes, over the whole area of the wall at regular intervals, varying in different examples between three and seven courses. Traces of the reeds can be seen in nearly all the great walls, and this precaution seems to have been deemed essential for the stability of a brick mass from the Old Kingdom down to the latest times, both in Upper Egypt and in the Delta. Still another precaution was sometimes taken to prevent cracking, which consisted of the insertion in the body of the brickwork of grids of wooden beams laid longitudinally and transversely through it about every fifth course.

In the Old and Middle Kingdoms the great walls were laid on level beds. In the New Kingdom, however, though precisely at what period we cannot determine,[1] the wall was divided up into comparatively short sections, one section being laid on a bed concave to the horizon, and the adjoining sections on convex or level beds (Figs. 252–5). Each section is—in theory at any rate—a separate entity, and the bricks of one do not course with those of another.[2] Where the original faces of the wall are preserved, they have the appearance of panelling, the faces of alternate sections being laid from a few inches to nearly two feet in front of the others.

[1] It has been considered that the brick wall (Fig. 254) bounding the area at Abydos known as the 'Kom el-Sultân' is Middle Kingdom. Although sculptures of this period have been found in the near vicinity, the date of the brick wall has yet to be definitely established.

[2] They are not found exclusively on alluvium. The wall of El-Kâb, for instance, is laid on the surface of the desert.

Fig. 252. The great brick wall of El-Kâb, constructed in sections on beds alternately convex and concave. (Photograph by N. de G. Davies, Esq.)

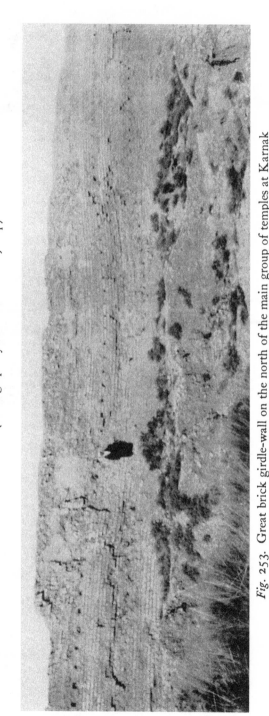

Fig. 253. Great brick girdle-wall on the north of the main group of temples at Karnak

Fig. 254. Brick girdle-wall at Abydôs (Kôm el-Sultân), showing sections laid alternately on concave and level beds

Fig. 255. Part of the outer face of the brick girdle-wall at Abydos (fig. 254), showing straight joint between wall and the gateway, and part of the original face of the wall

Fig. 256. Roman house-wall, 15 feet high, at Kôm Washîm, constructed on a concave bed. Probably fifth century A.D. (Photograph by J. Starkey, Esq.)

In the interior of the walls the bricks are almost always laid all headers, but the faces are laid in alternate courses of headers and stretchers—a method now known as English Bond (Fig. 255). The object must have been to attempt to make the courses of headers inside the wall break joint with those above and below them, and thus to avoid the chance of a longitudinal split in the wall. As an additional safeguard, however, the layers of rushes or matting over the bedding joints at regular intervals were almost invariably employed.

It has been supposed by some students that the late brick walls were constructed by first building the towers on concave beds and then filling in the space between them with brickwork on convex or level beds, or vice versa. In many walls, however, it is not uncommon to find several consecutive sections without any straight joint dividing them. This can be very clearly seen in the wall of El-Kâb, where a line of bricks of a slightly different colour from the rest runs through several sections continuously. In the wall at Armant the size of the bricks abruptly changes when a level of about five feet above the ground is reached, this change occurring throughout the sections of which the wall is composed. This would hardly happen had the sections been built at different times. Another reason for supposing that the sections were sometimes built simultaneously is that the longitudinal beams inserted in the mass of the brickwork not infrequently pass from one section to another. It does not follow, however, that the early examples of walls of this type were not constructed in alternate sections to facilitate drying; indeed it is very likely that the peculiarities of the examples mentioned, all of which are of late date, are due to ignorance on the part of the builders of the original purpose of the sections, and are a blind continuation of a type which had, as it were, become standardized. Some students, on the other hand, are of opinion that the idea underlying the practice of constructing a wall in sections was to prevent longitudinal contraction and the breaking up of the mass.

The custom of building brick walls on curved beds was continued down to Roman times, and examples are found, not only in town walls, but in the larger houses as well (Fig. 256). The modern Egyptian, however, never builds his mud-brick walls otherwise than on level beds, and with level courses throughout.

In the construction of brick pyramids, such as those of El-Lahûn, Hawâra, and Dahshûr, many of the precautions described above are taken. The drying of the interior would be an even slower process than that of one of the great walls. The general method underlying the construction

of a brick pyramid seems to have been to give it a framework of fairly large
brick walls, two running at right angles to form the diagonals, and others,

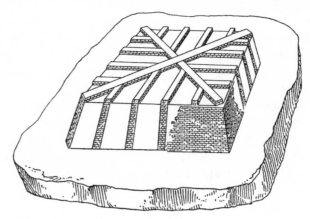

Fig. 257. Probable internal structure of the brick pyramid
at El-Lahûn. (After PERROT & CHIPIEZ, *Histoire de L'Art
dans l'Antiquité*, i, p. 211.)

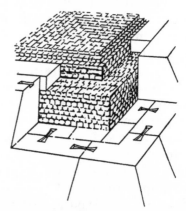

Fig. 258. Stone facing to a brick pyramid at Dahshûr. (From DE MORGAN,
Fouilles à Dahchour, Mars–juin, 1894, p. 48.)

not bonded into the diagonals, running parallel to its sides. The structure
was very likely completed, after the framework had been allowed to dry, by
laying brickwork and filling with brick-earth between the walls already laid
(Fig. 257). In certain brick pyramids stone was employed to strengthen
the diagonal walls; great blocks of limestone can be seen used for this pur-

Fig. 259. Remains of gateway in the Old Kingdom brick fortress (?) at Abydos. ('The Shûna'), showing traces of the plastered panels with which it was faced

pose at the corners of some of the courses of the pyramid of El-Lahûn. In the few cases where any of the stone facing remains on the brick pyramids (Fig. 258), the brickwork behind it is laid in the form of steps. This was almost certainly constructed last, and arranged to suit the different heights of the stone courses.

Two curious forms of brickwork remain to be noticed: the 'wavy walls', and the brick panelling used on the walls of mastabas and occasionally, in early times, on the great enclosure walls.

The Egyptians discovered early that if they wished to construct a wall only half a brick in thickness, a good method of doing so was to make it

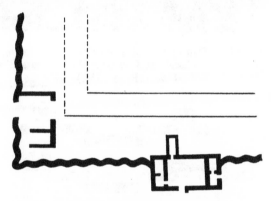

Fig. 260. Undulating brick girdle-wall, 41½ inches thick, round the Middle Kingdom pyramid at Mazghûna. (After Petrie, *The Labyrinth, Gerzeh and Mazghuneh*, Pl. XXXIX.)

run, not in a straight line, but undulating to and fro along its length, a complete undulation taking place every six bricks' length or so. Sundry walls of this type have been found used to keep back the sand during the excavation of tombs. In the Middle Kingdom walls of considerable size were built in this manner, a good example being seen in the temenos wall of the destroyed pyramid at Mazghûna, which is 41½ inches thick (Fig. 260). Others are known at Abydos, and at Koshtemna in Nubia.

In the Old Kingdom the walls of the mastabas, and even those of some of the brick forts, as at Hieraconpolis and Abydos (Fig. 259), have their exterior faces constructed in the form of panelling, and even in the earliest times this is sometimes of considerable intricacy (Fig. 261). Since this effect is reproduced on many of the Old Kingdom and Middle Kingdom sarcophagi, which are obviously meant to imitate houses, we must look for some feature in very early houses to which it may owe its origin. In some of the graves in the protodynastic cemeteries at Tarkhân, the tomb-chambers had been covered in, and sometimes lined with wooden planks, a fair proportion of which had been perforated at various points and in

Fig. 261. Plan of part of a brick wall surrounding a protodynastic mastaba at Tarkhân. (After PETRIE, *Tar-khan*, i, Pl. XVIII.)

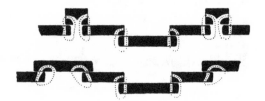

Fig. 262. Section of planks found in the proto-dynastic cemetery of Tarkhân, showing how they may have been bound together for use as walls of portable houses. (After PETRIE, *Tar-khan*, i, Pl. IX.)

various ways along their edges. Prof. Petrie, after a detailed examination of these planks, came to the conclusion that they had been taken from ancient wooden houses, which had been more or less portable, and which had been, as it were, laced together by thongs. Planks so joined would form a house having very much the appearance of being panelled (Fig. 261). The writers are not aware how far this explanation is generally accepted; at any rate, no better reason has yet been advanced for the slots in these ancient planks.

In Egypt, as in other countries, bricks were now and then made of special forms for special purposes, but space forbids a list of the various 'sport' bricks which have been found. Two examples, however, are of interest.

In the brick pyramids which were placed above the XIXth dynasty tombs
at Dirâ' Abu'l Naga at Thebes, the outside bricks of the course immediately
below the pyramid proper have their edges moulded into the form of a
cornice. Bricks in the form of a quadrant of a circle are also known in the
houses of the XVIIIth dynasty at El-'Amârna, which were used in the
construction of columns.

XX

EGYPTIAN MATHEMATICS

SINCE mathematics of a sort must play a part in all constructional work of any magnitude, as, for example, in the determination of the areas and proportions of buildings and the calculation of the batter on pyramids and walls, it may not be out of place, before concluding the present survey of the Egyptians' building methods, to attempt briefly to review what is at present known of their manner of calculating.

Our knowledge is derived from the following sources:

(*a*) A papyrus dated to the time of King 'Auserrē'-Apepa, a Hyksos king, and purporting to have been copied from a document of the time of King Amenemhēt III, and now known as the 'Rhind Papyrus'.[1]

(*b*) A document now known as the 'Moscow Papyrus', of the Middle Kingdom; still partly unpublished.[2]

(*c*) Portions of XIIth dynasty papyri, known as the 'Kahûn Fragments'.[3]

(*d*) A Middle Kingdom papyrus, now in Berlin.[4]

(*e*) Two wooden tablets, of the Middle Kingdom, in the Cairo Museum (Cat. no. 25367/8).[5]

(*f*) A Demotic papyrus of the Roman Period, containing tables of fractions.[6]

(*g*) Several tables of fractions of Byzantine date.[7]

(*h*) A Coptic ostrakon with tables of fractions.[8]

(*i*) The 'Mathematical Papyrus from Akhmim', dating to between the sixth and the ninth centuries A. D.[9]

To these must be added a long satirical letter known as *Anastasi Papyrus I*, which has already been quoted (p. 92), and which, though it gives no working out of problems, shows that the Egyptians were able

[1] PEET, *The Rhind Mathematical Papyrus.*

[2] TURAIEV, *Ancient Egypt*, 1917, pp. 100–2. Also see footnote on p. 223.

[3] GRIFFITH, *Hieratic Papyri from Kahun and Gurob*, pp. 15–18.

[4] SCHACK-SCHACKENBURG, *Aegyptische Zeitschrift*, 38, 135 ff.

[5] PEET, *Journal of Egyptian Archaeology*, ix, pp. 91 ff.

[6] REVILLOUT, *Mélanges sur la métrologie, et l'économie politique de l'ancienne Égypte*, and HULTSCH, *Neue Beiträge zur aegyptischen Teilungsrechnung* (*Bibliotheca Mathematica*, ser. 3, vol. ii, pp. 177–84).

[7] THOMPSON, *Ancient Egypt*, 1914, p. 52.

[8] SETHE, *Von Zahlen und Zahlworten bei den alten Aegyptern*, 70, 71, and CRUM, *Coptic Ostraka*, no. 480.

[9] BAILLET, *Le Papyrus mathématique d'Akhmim* (*Mém. Miss. franç. du Caire*), vol. 9, fasc. 1.

to calculate the number of bricks required for a sloping embankment with a batter and with internal compartments, and to determine the weight of an obelisk.

Of the above documents *The Rhind Mathematical Papyrus* is the most important, from the point of view both of completeness and of the variety of subjects treated.[1]

Before the nature of the Egyptians' mathematics can be properly appreciated, their peculiar mentality has to be borne in mind. This has been aptly expressed in the following terms: 'Despite the reputation for philosophic wisdom attributed to the Egyptians by the Greeks, no people has ever shown itself more averse to speculation or more whole-heartedly devoted to material interests.'[2] This statement applies even more forcibly to their mathematics than to the other branches of their science. Prof. Peet sums up the mentality of the Egyptian as regards mathematics by saying that he 'does not speak or think of 8 as an abstract number, he thinks of 8 loaves or 8 sheep. He does not work out the slope of the sides of a pyramid because it interests him to know it, but because he needs a practical working rule to give to the mason who is to dress the stones. If he resolves $\frac{2}{13}$ into $\frac{1}{8} + \frac{1}{52} + \frac{1}{104}$, it is not because this fact in itself appeals to him in any way, but simply because sooner or later he will come across the fraction $\frac{2}{13}$ in a sum, and since he has no machinery for dealing with fractions whose numerators are greater than unity, he will then urgently need the resolution above stated'.[3]

The Egyptian system of notation was decimal. There was a sign each for unity, for ten, for one-hundred, and for the higher powers of ten as far as a million. Integers were expressed by writing each sign as often as it was required. It will be seen that with this system it would require twenty-four hieroglyphic signs to write the number 879, though in the cursive script many abbreviations were used.

It was in the manner in which fractions were expressed that the Egyptian system of notation shows such peculiar features, rendering their method of calculation so very different from ours. A fraction was represented by writing the mouth-sign, probably reading *ro* and meaning 'a part', above the number which we should describe as the denominator. In Egyptian,

[1] Prof. Peet's work on the Rhind papyrus (*op. cit.*) contains all the student requires to know in the first instance of the other documents, and should be in the hands of any one desiring more than a most superficial knowledge of Egyptian mathematics. Apart from some help received from Mr. Battis-

combe Gunn, and from some notes of Prof. Alan Gardiner, nearly all the information in the present chapter has been taken from Prof. Peet's publication.
[2] GARDINER, *Egyptian Grammar*, p. 4.
[3] PEET, *op. cit.*, p. 10.

the number following the word *ro* had an ordinal meaning, and $\stackrel{ro}{5}$ means 'part five' or the 'fifth part', which concludes a row of equal parts together constituting a single set of five. As being the part which concluded the row into one series of the number indicated, the Egyptian *ro*-fraction was necessarily a fraction with, as we should say, unity as the numerator. To the Egyptian mind it would have seemed nonsense and self-contradictory to write *ro* $\frac{4}{7}$, for in any series of seven, only one part could be the seventh, namely, that which occupied the seventh place in a row of seven equal parts laid out for inspection.[1] The Egyptian, though he must have been able to realize, for example, what four parts out of seven meant, had to express $\frac{4}{7}$ as $\frac{1}{2}+\frac{1}{14}$; similarly, $\frac{11}{49}$ had to be expressed as $\frac{1}{7}+\frac{1}{14}+\frac{1}{98}$, or by other aliquot fractions giving the same sum.

In addition to the *ro*-sign, the Egyptian had one to express $\frac{1}{2}$ and, curiously enough, another representing $\frac{2}{3}$, which actually meant 'the two parts' (out of three), and still another for $\frac{3}{4}$. Though the sign for $\frac{2}{3}$ plays an important part in Egyptian arithmetic, that for $\frac{3}{4}$ does not appear to be used at all, having only a metrological use.

The most important computations for the Egyptian were those in which multiplication and division entered, the latter operation often presenting considerable difficulties.

In order to carry out division processes more rapidly, the Egyptian had worked out the values, expressed as a series of *ro*-fractions, of the number 2 divided by all the odd numbers up to 101, and these the scribe used very much as we use our logarithm-tables. There seem to have been other tables giving the value of $\frac{2}{3}$ of numbers, though what their range may have been is not known. It appears that it was not until very late times that the expression of other divisions, such as $1 \div 7$, $2 \div 7$,[2] $3 \div 7$, &c., as a sum of aliquot parts, was set out in the form of tables.[3]

The resolutions of 2 divided by the odd numbers from 5 to 47 were:

$2 \div 5$	$\frac{1}{3} + \frac{1}{15}$	$2 \div 17$ $\frac{1}{12} + \frac{1}{51} + \frac{1}{68}$
$2 \div 7$	$\frac{1}{4} + \frac{1}{28}$	$2 \div 19$ $\frac{1}{12} + \frac{1}{76} + \frac{1}{114}$
$2 \div 9$	$\frac{1}{6} + \frac{1}{18}$	$2 \div 21$ $\frac{1}{14} + \frac{1}{42}$
$2 \div 11$	$\frac{1}{6} + \frac{1}{66}$	$2 \div 23$ $\frac{1}{12} + \frac{1}{276}$
$2 \div 13$	$\frac{1}{8} + \frac{1}{52} + \frac{1}{104}$	$2 \div 25$ $\frac{1}{15} + \frac{1}{75}$
$2 \div 15$	$\frac{1}{10} + \frac{1}{30}$	$2 \div 27$ $\frac{1}{18} + \frac{1}{54}$

---

[1] GARDINER, *Egyptian Grammar*, p. 196, but see also p. 222 of this volume.

[2] The Demotic papyrus (p. 216) contains part of such a list.

[3] In the Rhind papyrus, the resolutions of 2 divided by the odd numbers are not given in the form shown above, but each is treated as a separate little theorem to prove its correctness.

$2 \div 29 \quad \frac{1}{24} + \frac{1}{58} + \frac{1}{174} + \frac{1}{232}$ $\qquad 2 \div 39 \quad \frac{1}{26} + \frac{1}{78}$

$2 \div 31 \quad \frac{1}{20} + \frac{1}{124} + \frac{1}{155}$ $\qquad 2 \div 41 \quad \frac{1}{24} + \frac{1}{246} + \frac{1}{328}$

$2 \div 33 \quad \frac{1}{22} + \frac{1}{66}$ $\qquad 2 \div 43 \quad \frac{1}{42} + \frac{1}{86} + \frac{1}{129} + \frac{1}{301}$

$2 \div 35 \quad \frac{1}{30} + \frac{1}{42}$ $\qquad 2 \div 45 \quad \frac{1}{30} + \frac{1}{90}$

$2 \div 37 \quad \frac{1}{24} + \frac{1}{111} + \frac{1}{296}$ $\qquad 2 \div 47 \quad \frac{1}{30} + \frac{1}{141} + \frac{1}{470}$

It is not always clear why the Egyptians selected these values for the division of 2 in preference to the alternative solutions of many of them. It seems that these tables were the result of the accumulated experience of many generations, the equivalents given having been found the most easy with which to work.

The Egyptian method of multiplication and division was little more than a system of trial and error carried out by doubling, halving, and multiplying by two-thirds. Two-thirds of a quantity was taken at one operation, and from it, if necessary one-third, one-sixth, etc. were found. The process of finding two-thirds of whole numbers offers no difficulties. With regard to fractions, the ancient method was to add the half to the sixth part. Thus $\frac{2}{3}$ of $\frac{1}{5}$ is $\frac{1}{10} + \frac{1}{30}$; likewise $\frac{2}{3}$ of $\frac{1}{11}$ was taken as $\frac{1}{22} + \frac{1}{66}$. Why one-third of a quantity was not taken first has not, as yet, been satisfactorily explained. Multiplying by a number greater than two (except 10) seems to have been very rarely performed. The Rhind papyrus, being a more or less advanced work, hardly gives any examples of simple multiplication or division. All show some peculiar complication, and nearly all omit some steps which were obvious to the Egyptian but not always so to the modern mind. Following are examples of a simple multiplication and division carried out according to the ancient method, which may enable the reader better to understand the literal translation of a more or less similar problem from the Rhind papyrus, in which several steps are omitted.

(1)   *Multiply* $1 (+)\frac{1}{2} (+)\frac{1}{4}$ by $23\frac{1}{3}$

            $\diagdown$ 1 (times $1 \frac{1}{2} \frac{1}{4} =$) $1 \frac{1}{2} \frac{1}{4}$

            $\diagdown$ 2   ,, $3 \frac{1}{2}$

            $\diagdown$ 4   ,,   7

              8   ,,   14

           $\diagdown$ 16   ,,   28

              $\frac{2}{3}$,, $\frac{2}{3} \frac{1}{3} \frac{1}{6}$

           $\diagdown$ $\frac{1}{3}$,, $\frac{1}{3} \frac{1}{6} \frac{1}{12}$

      Total[1] $23\frac{1}{3}$ *Answer* $40 \frac{1}{2} \frac{1}{3}$

(The addition of the fractions is discussed on p. 221.)

[1] Of items marked with a tick.

(2)   *Summon* 49 *from* 11 (divide 49 by 11)

$$1 \text{ (multiplied by 11 makes) } 11$$
$$2 \qquad \text{,,} \qquad 22$$
$$\backslash 4 \qquad \text{,,} \qquad 44$$

Total 4   *Remainder* 5

Two summoned from 11 *is* $\frac{1}{6} \frac{1}{66}$ *Found* (p. 218).

Treat 2 *to find* 5

$$1 \text{ (times 2 is) } 2$$
$$\backslash 2 \qquad \text{,,} \qquad 4$$
$$\backslash \tfrac{1}{2} \qquad \text{,,} \qquad 1$$

Answer $2\frac{1}{2}$

Multiply $\frac{1}{6} \frac{1}{66}$ *by* $2\frac{1}{2}$

$$1 \ \left(times\ \tfrac{1}{6} \tfrac{1}{66}\ is\right)\ \tfrac{1}{6}\tfrac{1}{66}$$
$$\backslash 2 \qquad \text{,,} \qquad \tfrac{1}{3}\tfrac{1}{33}$$
$$\backslash \tfrac{1}{2} \qquad \text{,,} \qquad \tfrac{1}{11} \text{ (because } \tfrac{1}{2} \text{ of } 2 \div 11 \text{ must be equal to } 1 \div 11)$$

Total $2\frac{1}{2}$ *Answer* $\frac{1}{3}\frac{1}{11}\frac{1}{33}$

Add the whole number 4

Final answer $4\frac{1}{3}\frac{1}{11}\frac{1}{33}$

It will be seen that the processes are (1) to ascertain how many complete times 11 goes into 49, and with what remainder, and (2), knowing the value of $2 \div 11$, to find, by multiplying it by $2\frac{1}{2}$, what is the value of the remainder, 5, divided by 11.

(3)   Rhind Problem no. 30

'If a scribe say to thee:
    " 10 *has become* $\frac{2}{3}\frac{1}{10}$ *of what?"*

Let him hear:

You are to treat $\frac{2}{3}\frac{1}{10}$ *to find* 10

$\backslash 1$	$\frac{2}{3}$	$\frac{1}{10}$
2	$1\frac{1}{3}$	$\frac{1}{5}$
$\backslash 4$	$3\frac{1}{15}$	
$\backslash 8$	$6\frac{1}{10}$	$\frac{1}{30}$

Total 13   (*Remainder*)[1] $\frac{1}{30}$

$\frac{1}{30}$ *is multiplied* 23 *times to find* $\frac{2}{3}\frac{1}{10}$
Total, this quantity that says it, $13\frac{1}{23}$

[1] The word 'Remainder' in this problem is omitted in the papyrus.

(Proof)

$$\begin{array}{ll}
1 & 13\frac{1}{23} \\
\backslash\frac{2}{3} & 8\frac{2}{3}\ \frac{1}{46}\ \frac{1}{138} \\
\backslash\frac{1}{10} & 1\frac{1}{5}\ \frac{1}{10}\ \frac{1}{230}
\end{array}$$

Total 10

It will be noticed that the process of the addition of the fractions and the determination of what fraction of $\frac{2}{3}\frac{1}{10}$ amounts to $\frac{1}{30}$ are not given. Other problems, however, give the full working for the addition of fractions, which differs little in principle from the modern method of reducing them all to a common denominator. In Rhind Problem no. 32, a series of fractions have to be added and proved to have the value of $\frac{1}{4}$, and the working is as follows:

$\frac{1}{12}$	$\frac{1}{114}$	$\frac{1}{228}$	$\frac{1}{18}$	$\frac{1}{36}$	$\frac{1}{342}$	$\frac{1}{684}$	$\frac{1}{24}$	$\frac{1}{48}$	$\frac{1}{456}$	$\frac{1}{912}$
76	8	4	$50\frac{2}{3}$	$25\frac{1}{3}$	$2\frac{2}{3}$	$1\frac{1}{3}$	38	19	2	1 [1]

Total 228 (i.e.) $\frac{1}{4}$

Prof. Peet explains the process by saying that 'all the fractions or aliquot parts seem to have been reduced to the terms of the highest aliquot part, namely the 912th part: under each fraction is placed, in red, the number of 912ths contained in that fraction, a number which, it will be observed, is not in all cases a whole number. The step must involve a certain amount of rough working, which is always omitted in the papyrus. The red figures are now added and come to 228, which is $\frac{1}{4}$ of 912. Therefore the sum of all these fractions is the required $\frac{1}{4}$. . . .' [2]

In Rhind Problem no. 30, cited above, the addition of the fractions was considerably simpler, and would have been worked out as follows:

9 added to	$\frac{2}{3}$	$\frac{1}{10}$	$\frac{1}{15}$	$\frac{1}{10}$	$\frac{1}{30}$
	20	3	2	3	1

Total (for the fractions) 29, giving a result for the sum of the fractions as $\frac{1}{30}$ short of unity, and for the whole sum that amount short of 10.

The second omitted step in the problem, namely the determination of how many times $\frac{1}{30}$ make $\frac{2}{3}\frac{1}{10}$, would have been by the normal method of division, and the working would have been:

$$\begin{array}{lll}
\backslash 1 \text{ (times } \frac{1}{30} \text{ is)} & \frac{1}{30} \\
\backslash 2 & \text{,,} & \frac{1}{15} \\
10 & \text{,,} & \frac{1}{3} \\
\backslash 20 & \text{,,} & \frac{2}{3} \\
\text{Total } 23 & & \frac{2}{3}+\frac{1}{15}+\frac{1}{30} \text{ or } \frac{2}{3}+\frac{1}{10}
\end{array}$$

[1] In the papyrus this row of numbers is in red.   [2] PEET, *op. cit.*, p. 17.

On the other hand, it is very possible that the Egyptian would at once recognize that 3 times $\frac{1}{30}$ is $\frac{1}{10}$, and his rough working would then have been:

$$
\begin{array}{ll}
\text{I} & \frac{1}{30} \\
\diagdown 3 & \frac{1}{10} \\
\text{I0} & \frac{1}{3} \\
\diagdown 20 & \frac{2}{3} \\
\text{Total } 23 & \frac{2}{3}+\frac{1}{10}
\end{array}
$$

Prof. Peet explains the step by saying[1] 'since $\frac{2}{3}\,\frac{1}{10}$ is $\frac{23}{30}$, a step which is completely omitted, the reply must be $\frac{1}{23}$'. To carry out this step by that chain of reasoning seems to the writer to imply that the Egyptian understood what $\frac{23}{30}$ meant, or as he might have thought of it, 23 parts out of 30, and this is strongly suggested by his method of adding fractions by means of a common denominator. Be this as it may, his notation was incapable of expressing such a fraction other than as a series of aliquot parts, helped out, if necessary, with the fraction $\frac{2}{3}$.

In the Rhind Papyrus a seemingly unnecessary number of examples is given of multiplication and division, which we, with algebra to help us, could cover with a few lines of explanation or by a formula. The reason for this is that each problem when worked out by the system of trial multiplications offers its peculiar difficulties, some of which take a considerable amount of ingenuity to solve.

Equations of the first degree are solved by simple method of trial, and equations of the second degree where there is virtually one unknown were also understood. In the Berlin papyrus (see p. 216) 100 square cubits have to be divided into two squares whose sides are to one another in the ratio of 1 to $\frac{3}{4}$.

The conception of squaring and extracting the square root was familiar to the Egyptians. While the former process is merely a question of multiplication, the reverse process for all but the simplest quantities must have involved a long series of trials. The Berlin Papyrus gives the solutions for the square root of $6\frac{1}{4}$ and $1\frac{1}{2}\frac{1}{16}$.

Although the ancient approximation of the ratio of the circumference of a circle to its diameter, or π, is not given in the mathematical papyri, the determination of the area of a circle occurs in the Rhind Papyrus (no. 50). The method was to subtract $\frac{1}{9}$ of the diameter from it and to square the result. This would now be expressed by the formula $A = (\frac{8}{9}\,D)^2$. This approximation is fairly close, the area thus obtained being $\cdot 7902\,D^2$ instead of the true value $\cdot 7854\,D^2$, and must have been originally obtained by drawing a circle on a finely squared surface and counting the squares.

[1] PEET, *op. cit.*, p. 65.

Though there is a fair amount of evidence that in late times triangular fields were measured for taxation by halving the product of the longest and shortest sides, there is little doubt that in the earlier periods the area was correctly known to be half the product of the base and the vertical height or *emrōyet*. Further it was known that the area was one-half that of the rectangle described on its base with the same vertical height.

The volume of a cylinder was found by multiplying the square of diameter minus its ninth part (see above) by the length, and that of a symmetrical pyramid (at least) was undoubtedly known to be one-third the base area multiplied by the height. How this was originally determined is not known, since to prove it mathematically requires knowledge beyond that which the Egyptians possessed, and to demonstrate it practically by cutting out a pyramid from a parallelopiped is a very complicated process. Mr. Battiscombe Gunn has suggested to me that it was arrived at by weighing a parallelopiped of clay or mud and then weighing the pyramid cut from it, a simple and practical method which the Egyptian would very likely have used.

One of the most surprising solutions obtained by the Egyptians was the determination of the volume of a truncated pyramid. If H is the vertical height, a the side of the square base, and b that of the square top, the formula for the volume is $\frac{H}{3} (a^2 + ab + b^2)$, which was the form in which it was known to the Egyptians. A truncated pyramid can be resolved into a parallelopiped, four wedges and four oblique pyramids which can be assembled to form one symmetrical pyramid. It can be shown[1] that, by a simple graphical process, the formula found by the breaking up of the figure can be converted into the more handy one used to-day by the simplest of graphical methods.

Enough has been brought forward to indicate the general characteristics of the Egyptian methods of computation. The knowledge of multiplication and division was amply sufficient to solve any problem which might be encountered in constructing a temple, pyramid, or wall and ascertaining the weight of the material used. How much further the Egyptian had progressed must remain unknown until further discoveries are forthcoming, but it is very unlikely that mathematics of a much more advanced nature were ever in the possession of the scribes. To obtain such it would have been necessary for the Egyptian to change, not only his system of notation, but also the nature of his mind.

[1] GUNN and PEET, *Four Geometrical Problems from the Moscow Mathematical Papyrus* (*Journal of Egyptian Archaeology*, xv, pp. 167–185).

APPENDIX I

ANCIENT EGYPTIAN TOOLS

THE tools shown in the illustrations (Fig. 89, p. 93, and Figs. 263–7) form a fairly complete collection of those known to have been used by the ancient architects, quarrymen, and masons. All are in the Cairo Museum with the exception of the mason's cord and reel, which is in the Metropolitan Museum, New York.

Among the tools not shown may be mentioned the cubit rod, the stonemason's adze, the brickmaker's hoe, and the 'plug and feathers' (p. 18). The cubit rod has been omitted, since a description suffices and since the Cairo Museum possesses but one example which might conceivably have been used by the masons, namely the very rough, ill-divided rod found in the tomb of Sennūtem at Thebes. The fine wooden specimens from the tomb of Tut'ankhamūn, though now on view in Cairo, are not yet available for description.

Though carpenter's adzes, and models of them, are well known, none has yet been found which is known to have been used in stone-dressing. The shape of the blade changes considerably during the course of Egyptian history, but the manner of its attachment to the haft remained practically the same (Fig. 36).

Many brickmakers' (or cultivators') hoes have been found. They are always of wood, the 'tang' of the blade fitting into a hole in the haft, the two members being retained at the desired angle by a lashing of palm-rope (Fig. 251). The metal-hafted hoe, called fâs or ṭûria, of the form used to-day in Egypt (Fig. 250), does not appear to date further back than Roman times.

Iron splitting-wedges and 'feathers' have been found in the Ramesseum,[1] but they are certainly of very late dynastic, if not of Roman date.

Photographs of Egyptian hammers have been omitted, as good examples are represented in the hands of the ancient boat-builders (Fig. 36). Hammers of modern form, with metal heads, have not been found of dynastic date, nor are they represented in the tomb-scenes.

Of the two tools which may well have been known, namely the mason's pick and the testing-plate, no example or illustration has been found. This also applies to the tubular drill, which was freely used.

The stonemason's wedge of wood (Fig. 267), probably of the Vth dynasty, is of considerable interest. It is stained with mud and bears unmistakable signs of crushing, and is fitted with a handle to prevent the fingers from being pinched when in use. This, and the wooden roller shown with it, have quite recently been found near the pyramid of Pepi II at Saqqâra, and are hitherto unpublished. They have been included here by kind permission of the Director-General of the Antiquities Department and of M. Gustave Jéquier, who discovered them.

[1] PETRIE, *Tools and Weapons*, Pl. XIII, B. 16, 17.

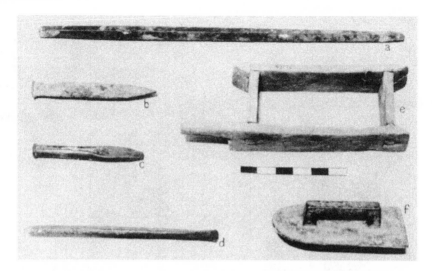

Fig. 263. *a,* long copper quarryman's chisel, from Gebelein; *b,* copper mason's chisel, New Kingdom, from Ghorâb; *c,* copper mason's chisel; *d,* copper mortise chisel; *e,* wooden brick mould, XVIIIth dynasty, from Thebes; *f,* copper plasterer's tool, date unknown. The scale is in inches. (All in Cairo Museum)

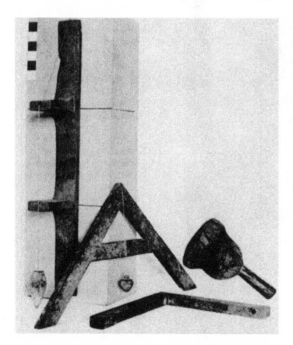

Fig. 264. Square, level, and plumb-rule, from the tomb of Sennûtem at Thebes; XXth dynasty. Mason's mallet from Saqqâra; Old Kingdom. Scale in inches. (Cairo Museum)

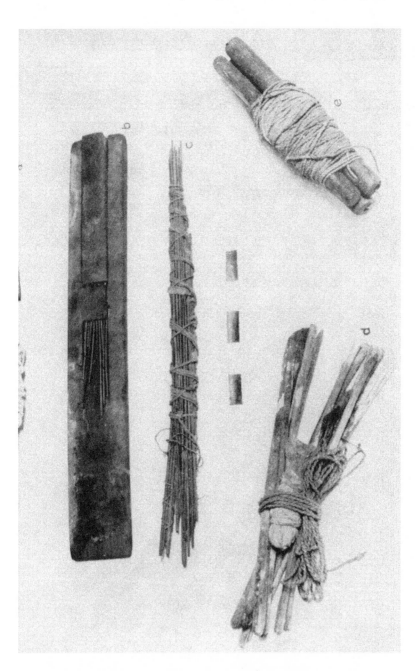

Fig. 265. *a,* mason's cord and reel (copy) from Thebes, XIth dynasty. Original in the Metropolitan Museum, New York; *b,* wooden palette with red and black ink and reeds, from Saqqâra; *c,* bundle of spare reeds; *d,* palm brushes for painting. These are tied with the ochre string used for drawing straight lines; *e,* set of 'boning-rods' from El-Qurna (Thebes), New Kingdom. Scale in inches. (All from Cairo Museum)

Fig. 266. Polishing tools (?) of black granite; date unknown, from Saqqâra. Pounding-ball of dolerite, from Aswân. Dolerite maul on haft; XIth dynasty, from Thebes. (Cairo Museum.) Scale in inches

Fig. 267. Wedge for handling stone and roller, probably of the Vth dynasty, from near the pyramid of Pepi II at Saqqâra. Both show clear traces of crushing. Scale in inches. (By permission of the Antiquities Dept. and the finder, M. Gustave Jéquier)

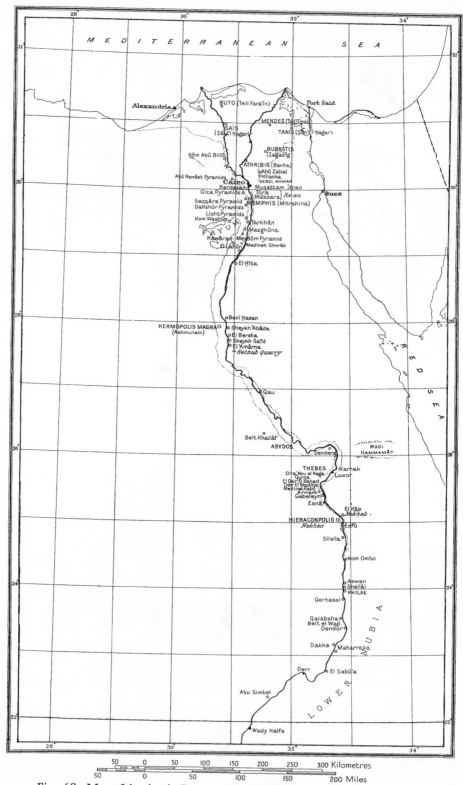

Fig. 268. Map of the sites in Egypt and Lower Nubia mentioned in the volume

APPENDIX II
LIST OF LOCALITIES IN EGYPT AND LOWER NUBIA
MENTIONED IN THE VOLUME

On the map, modern names are printed in plain type, classical names in capitals, and ancient Egyptian names in italics.

In the list below, the positions of the sites are given to the nearest degree of latitude. Those in brackets are too far south to be shown on the map.

The spelling of the modern Arabic names is that adopted by the Survey of Egypt, except that in the text the diacritical marks distinguishing the palatalized letters have been omitted, and all accents except the circumflex indicating a vowel *pronounced* long and carrying the stress.

Place	Lat.	Place	Lat.
Abu Rawâsh	30	El-Kâb	25
Abu Simbel	22	El-Lahûn	29
Abu Za'bal	30	El-Qurna	26
Abydos	26	El-Sabû'a	23
Alexandria	31	Erment (See Armant)	
'Amârna (See El-'Amârna)		Esna	25
Armant	25		
Ashmunein (El-)	28	Fayyûm (El-)	29
Aswân	24		
Athribis	30	Gebel Ahmar	30
'Ayan	30	(Gebel Barkal	18)
		Gebelein	25
Beit el-Wâli	24	Gebel Silsila	25
Beit Khallâf	26	Ghorâb (Medînet Ghorâb)	29
Benha	30	Gîza (El-)	30
Beni Hasan	28	Gurna (See El-Qurna)	
Bubastis	31	Gurob (See Ghorâb)	
Buto	31		
		Hermopolis Magna	28
Cairo	30	Het-Nub	28
		Hîbeh (See El-Hîba)	
Dahshûr	30	Hieraconpolis	25
Dakka	23		
Deir el-Bahari (See El-Deir el-Bahari)		Illahun (See El-Lahûn)	
Dendera	26	Ipsamboul (See Abu Simbel)	
Dendûr	23		
Derr	23	Kalabsha (See Qalabsha)	
Dirâ' Abu'l Naga	26	Karnak (El-)	26
		Kerdâsa	30
Edfu	25	Kertassi (See Qertassi)	
El-'Amârna	28	Khânka (El-)	30
El-Bersha	28	Kôm Abu Billo	30
El-Deir el-Bahari	26	Kôm Ombo	24
El-Hîba	29	Kurna (See El-Qurna)	

Place.	Lat.
Lahun (See El-Lahûn)	
Lisht (El-)	30
Luxor	26
Ma'sara	30
Maharraka	23
Mazghûna	29
Medînet Habu	26
Memphis	30
Mendes	31
Meydûm	29
Mitrahîna	30
Muqattam	30
Nekheb	25
Nekhen	25
Philae	24
Port Sa'îd	31
Qalabsha	24
Qâu	27
Qertassi	24
Qurna (See El-Qurna)	
Re-au	30

Place.	Lat.
Sâ el-Hagar	31
Sabû'a (See El-Sabû'a)	
Sais	31
Sakkara (See Saqqâra)	
Sân el-Hagar	31
Saqqâra	30
(Sedenga	20)
(Sesebi	20)
Shellâl (El-)	24
Sheykh 'Abâda	28
Sheykh Sa'îd	28
Silsila (Gebel)	25
(Soleb	20)
Suez	30
Tanis	31
Tarkhân	29
Tell el-Amarna (See El-'Amârna)	
Tell Fara'în	31
Tell Tmai	31
Thebes	26
Tura	30
Wady Halfa	22
Wady Hammamât	26
Zagazig (El-Zaqazîq)	31

APPENDIX III

CHRONOLOGY

ALTHOUGH the dates of the kings from the XVIIIth dynasty onwards are known with very fair accuracy, there is still a difference of opinion among scholars as to the dates of those previous to that period, which depends on whether the time which elapsed between the XIIth and the XVIIIth dynasties was long or short.[1] Since the short dating is by far the more generally accepted, the writers have adopted it in the following list and have followed, up to and including the XXVth dynasty, the chronology used by Dr. J. H. Breasted.

In the following list, only the names of the kings mentioned in the volume are given. The number before a king's name shows his place in the dynasty.

DYNASTY I (8 Kings)

No. 5 Den (Udymu?)
6 'Az-ieb
7 Smerkhet

c. 3400–2980 B.C.

DYNASTY II (9 Kings)

8 Kha'-sekhemui

DYNASTY III (9 Kings). *c.* 2980–2900 B.C.

1 Sa-nakht
3 Zoser
9 Sneferu

DYNASTY IV (8 Kings). *c.* 2900–2750 B.C.

1 Khufu (Cheops)
3 Kha'frē' (Chephren)
4 Menkewrē' (Mycerinus)

DYNASTY V (10 Kings). *c.* 2750–2625 B.C.

2 Sahurē'
3 Neferirkerē'
6 Newoserrē'
10 Unas

[1] The student who desires to study the merits of the two systems should consult the following works: BREASTED, *Ancient Records,* i. 38–75; PETRIE, *Historical Studies,* ii; MEYER, *Aegyptische Chronologie* (Königliche Preussische Akademie, 1904).

DYNASTY VI (6 Kings).   *c.* 2625–2475 B.C.

3 Pepi I
4 Mernerē'
5 Pepi II

DYNASTIES VII—X.   *c.* 2475–2160 B.C.

First Intermediate Period

DYNASTY XI (7 Kings).   *c.* 2160–2000 B.C.

7 Menthuhotpe IV

DYNASTY XII (8 Kings).   *c.* 2000–1788 B.C.

5 Senusret III (Sesostris) 1887–1849
6 Amenemhēt III (Amenemmes)   1849–1801

DYNASTIES XIII—XVII.   *c.* 1788–1580 B.C.

Second Intermediate Period, which includes the Hyksos

DYNASTY XVIII (12 Kings and 1 Queen).   *c.* 1580–1350 B.C.

1 Amasis I ('Ahmose)   1580–1557
2 Amenophis I   .   .   .   .   .   . ⎫
3 Tuthmosis I ('Thothmes')   .   .   .   .   . ⎬ 1557–1501
4 Tuthmosis III, including Tuthmosis II and Queen Hatshepsowet   1501–1447
8 Tuthmosis IV   1420–1411
9 Amenophis III   1411–1375
10 Amenophis IV = Akhenaten   .   .   .   . ⎫
11 Smenkhkerē'   .   .   .   .   .   . ⎬ 1375–1350
12 Tut'ankhaten = Tut'ankhamūn   .   .   .   . ⎪
13 Ay (Eye)   .   .   .   .   .   . ⎭

DYNASTY XIX (8 Kings).   *c.* 1350–1205 B.C.

1 Haremhab   1350–1315
2 Ramesses I 1315–1314
3 Seti I 1313–1292
4 Ramesses II   1292–1225
5 Meneptah 1225–1215
8 Seti II   1209–1205

DYNASTY XX (11 Kings).   *c.* 1200–1090 B.C.

2 Ramesses III   1198–1167
3 Ramesses IV   1167–1161
8 Ramesses IX   1142–1123

DYNASTY XXI (10 Kings).   *c.* 1090–945 B.C.

1 Nesibenebded (Smendes) 1090–1085

DYNASTY XXII (9 Kings).   *c.* 945–745 B.C.

1 Sheshonq I (Sesonchis, SHISHAK) 945–924
2 Osorkon I 924–895

DYNASTY XXV (3 Kings).   *c.* 712–663 B.C.

3 Taharqa (Taracos, TIRHAKAH) 688–663

DYNASTY XXIX (4 Kings).   *c.* 398–379 B.C.

2 Hakor (Achoris) 393–380

DYNASTY XXX (3 Kings).   *c.* 378–340 B.C.

3 Nectanebos II (Nekht-nebf) 358–340

PTOLEMAIC PERIOD.   *c.* 332–57 B.C.

2 Ptolemy II, Philadelphus 287–246
3 Ptolemy III, Euergetes 246–222

ROMAN PERIOD 30 B.C.—A.D. 378

2 Nero . . . . A.D. 54–68
6 Galba 68–69
8 Vespasian 69–79
10 Domitian 81–96
46 Diocletian 284–305

APPENDIX IV

LIST OF WORKS QUOTED

BAILLET. Le Papyrus mathématique d'Akhmim (Mém. Miss. franç. du Caire. 1892).

BORCHARDT. Das Grabdenkmal des Königs Neferirkerē'. (J. C. Hinrichs, Leipzig. 1909.)

Das Grabdenkmal des Königs Newoserrē'. (Hinrichs, Leipzig. 1907.)

Das Grabdenkmal des Königs Sahurē'. (Hinrichs, Leipzig. 1913.)

Das Re-Heiligtum des Königs Neuserrē'. (Hinrichs, Leipzig. 1907.)

Gegen die Zahlenmystik an der grossen Pyramide bei Gise. (Behrend & Co., Berlin. 1922.)

Längen und Richtungen der vier Grundkanten der grossen Pyramide bei Gise. (Springer, Berlin. 1926.)

BREASTED. Ancient Records. (5 vols., Camb. Univ. Press.)

Development of Religion and Thought in Ancient Egypt. (Chas. Scribner & Sons, N.Y. 1912.)

BRUNTON. Lahun I, The Treasure. (Bernard Quaritch, London. 1920.)

BOREUX. Étude de Nautique Égyptienne. (Mém. Inst. Français, Cairo. 1926.)

BUDGE. Egyptian Sculptures in the British Museum.

CARTER & MACE. The Tomb of Tut·ankh·amen. (2 vols., Cassell, London. 1923 and 1927.)

CHOISY. L'Art de Bâtir chez les Égyptiens. (Gauthier Villars, Paris. 1903.)

COLE. Determination of the exact size and orientation of the Great Pyramid. (Survey of Egypt, paper no. 39, Cairo, Govt. Press. 1925.)

COLIN CAMPBELL. Two Theban Princesses. (Oliver & Boyd, London. 1910.)

DAVIES. El Amarna. (6 vols., Bernard Quaritch, London. 1903–8.)

The Tombs of Two Officials of Tuthmosis IV. (Egypt Expl. Society, London. 1923.)

DE MORGAN. Fouilles à Dahchour. (2 vols., Holzhausen, Vienna. (1) Mars–juin 1894; (2) 1895.)

ENGELBACH. The Problem of the Obelisks. (Fisher Unwin, London. 1923.)

Riqqeh and Memphis VI. (Brit. School of Arch. in Egypt, London. 1914.)

FISHER. Giza. The Minor Cemetery. (Univ. Press, Philadelphia. 1924.)

GARDINER. Egyptian Hieratic Texts. (J. C. Hinrichs, Leipzig. 1911.)

Egyptian Grammar. (Clarendon Press, Oxford. 1927.)

GARSTANG. Mahâsna and Bêt Khallâf. (Bernard Quaritch, London. 1902.)

GAUTIER & JÉQUIER. Fouilles de Licht. (Ernest Leroux, Paris. 1896.)

GRIFFITH. Hieratic Papyri from Kahun and Gurob. (Bernard Quaritch, London. 1898.)

HÖLSCHER. Das Grabdenkmal des Königs Chephren. (Hinrichs, Leipzig. 1912.)

JÉQUIER. Les Éléments de l'Architecture Égyptienne. (Auguste Picard, Paris. 1924.)

Les Temples Ptolémaïques et Romains. (Albert Morancé, Paris. 1926.)

Les Temples Raméssides et Saïtes. (Albert Morancé, Paris. 1926.)

LAYARD. The Monuments of Nineveh and Babylon. (J. Murray, London. 1853.)

LEBAS. L'Obélisque de Louxor. (Paris. 1839.)

LEPSIUS. Denkmäler. (Nicolaische Buchhandlung, Berlin.)

Auswahl. (Georg Wigand, Leipzig. 1842.)

LUCAS. Ancient Egyptian Materials. (Edward Arnold, London. 1926.)

MEYER. Aegyptische Chronologie. (Königliche Preussische Akademie. 1904.)

NAVILLE. The Eleventh Dynasty Temple of Deir el-Bahari. (Egypt Expl. Fund. 1907, 1910, 1913.)

The Temple of Deir el-Bahari. (Egypt. Expl. Fund. 1907, 1910, 1913.)

NEWBERRY. Beni Hasan. (2 vols., Egypt. Expl. Fund. 1893.)

The Life of Rekhmara. (Constable, London. 1900.)

El Bersheh. (2 vols., Egypt. Expl. Fund. 1893.)

PERRING. The Pyramids of Giza. (3 vols., James Frazer, London. 1840–2.)

PERROT & CHIPIEZ. Histoire de l'Art dans l'Antiquité. (Hachette, Paris. 1882.)

PEET. The Rhind Mathematical Papyrus. (Hodder & Stoughton, London. 1923.)

PETRIE. A History of Egypt; vol. I. (Tenth edition, revised, Methuen, London. 1923.)

Gizeh and Rifeh.

Historical Studies II. (Brit. School of Arch. in Egypt, London. 1911.)

Illahun Kahun & Gurob. (David Nutt, London. 1891.)

Inductive Metrology. (Hargraves, London. 1877.)

Kahun Gurob and Hawara. (Kegan Paul, London. 1890.)

Lahun II. (Bernard Quaritch, London. 1923.)

Medum. (David Nutt, London. 1892.)

On the mechanical methods of the Ancient Egyptians. (Journal of the Anthropological Inst., Aug. 1882.) 2 vols.

Royal Tombs. (2 vols., Egypt. Expl. Fund. 1900 and 1901.)

Tarkhan I and Memphis V. (Bernard Quaritch, London. 1913).

The Labyrinth, Gerzeh and Mazghuneh. (Bernard Quaritch, London. 1912.)

The Palace of Apries and Memphis II. (Brit. School of Arch. in Egypt. 1909.)

PETRIE.   The Pyramids and Temples of Gizeh. (New ed., Leadenhall Press, London. 1885.)
Tools and Weapons. (Brit. School of Arch. in Egypt, London. 1917.)
Weights and Measures. (Brit. School of Arch. in Egypt, London. 1926.)

QUIBELL.   Hierakonpolis I. (Bernard Quaritch, London. 1900.)

QUIBELL & GREEN. Hierakonpolis II. (Bernard Quaritch, London. 1902.)

REISNER.   Models of Ships and Boats. (Cat. Gén. du Musée Égyptien, Cairo.)
The Early Dynastic Cemeteries at Naga ed Deir. (Hinrichs, Leipzig. 1908.)

ROCHEMENTEIX & CHASSINAT. Le Temple d'Edfou. (Mém. Miss. Arch. Français. 1892, 1897, 1918.)

SCHÄFER.   Ein Bruchstück altägyptischer Annalen.

SETHE.   Von Zahlen und Zahlworten bei den alten Aegyptern. (Trübner, Strassburg. 1916.)

TORR.   Ancient Ships. (Camb. Univ. Press.)

TYLOR & GRIFFITH. The Tomb of Paheri at El Kab. (Egypt. Expl. Fund. 1894.)

PERIODICALS

Ancient Egypt. (British School of Archaeology in Egypt. Macmillan, London.)
Annales du Service des Antiquités. (Antiquities Department, Cairo.)
Bulletin of the Metropolitan Museum of Art, New York.
Journal of Egyptian Archaeology. (Egypt. Exploration Society, London.)
Zeitschrift für Aegyptische Sprache. (Hinrichs, Leipzig.)

INDEX

Masonry, oblique joints in ancient, 100, 106.
— on concave beds, 116.
— small-block, of the IIIrd dynasty, 8, 96.
— 'Type A' and 'Type B', 101, 103.
— unfinished, 86, 125, 145, 192, 195.
— with blocks of unequal height, 107.
Masons, ancient scene of, 105, 198.
Mason's pick, 17, 31, 194, 202, 224.
Mason's square, 96, 100, 224.
Mastabas, angle of, casing of, 118.
— the origin of pyramids, 6, 118.
— top of primitive, 6.
Masts, 42.
Mathematics, ancient Egyptian treatises on, 216.
— characteristics of ancient Egyptian, 217.
Measurement, of quarry-work, 21, 29.
— units of, 63–5.
Medînet Habu, Palace of Ramesses III, windows of, 173.
— Temple of Ramesses III, arch in, 185.
— — doorway in, 165.
— — roof-slabs in, 155.
— Temple of Tuthmosis III, doorway in, 165.
— — roof drainage in, 157.
— — window in, 171.
— XXVth dynasty shrines, arches in, 185.
Meketrē', model boat of, 42.
— model kiosk of, 173.
Memphis, 6, 69, 208.
Mendes, 14, 208.
Menkewrē', King, log roofs in temple of, 9.
Menthuhotpe IV, quarrying expedition under, 32.
Mereruka (Mera), 'false-door' of, 163.
Merirē', tomb of, at El-'Amârna, 54–6.
Mernerē', King, quarrying inscription of, 21.
Metal of ancient tools, 18, 24, 25.
Meydûm, pyramid of (see Pyramid of Sneferu).
Miners' map on papyrus, 56.
Models illustrating ancient method of building, 107.
— illustrating ancient method of dressing blocks, 102.
— scale, 59.
Modern granite quarrying, 25.
Monolithic column, erection of, 148.
— shaping of, 146.
Monolithic quartzite chamber, 23, 34.
Mortar, 78–83, 110, 113, 155, 197.
Mosaic of polygonal blocks on temple roofs, 157.
Moulding, Torus, origin of, 6.
— shaping of, 197.
Mullions, 173.
Multiplication, ancient method of performing, 219.
Mycerinus (see Menkewrē').

Natron, 200.
Nectanebos II, gateway of, at Karnak, 86, 190.
Nesibanebded (Smendes), quarry inscription of King, 21.
Numeral-system of the ancient Egyptians, 217.

Obelisk, barge carrying, 34, 39, 41, 89.
— quarrying an, 27.
— transport of, under Ptolemy Philadelphus, 35.
Oblique joints in masonry, 100, 106.
Obsidian, cutting of, 202 (Fig. 245).
Ochre, 16, 27, 29, 48, 199, 224.
Octagonal columns, 132, 136.
Offering table, scene showing shaping of, 198.
Open quarries, 13.
Orpiment used in pigments, 200.
'Osireion' (see Abydos).
Ostraka, 51.
Outline-draughtsman, 199, 201.

Packing-blocks in pyramids, 104, 109, 122.
Paddles, steering, 40.
Painting, imitative, 200.
Paints, composition of ancient, 200.
Palace of Ramesses III (see Medînet Habu).
Pall shown in ancient tomb-plan, 50.
Palm, unit of length, 50, 52, 56, 63.
Palm columns, 144.
Palm-frond capitals, 6, 160.
Palm-fronds possibly used in constructing early huts, 6.
Panelling in brick, 5, 210, 213.
— in stone, 5, 213.
Papyrus columns, 140.
Papyrus, used in constructing early huts, 6.
Patches in masonry, 99, 151.
'Patchwork masonry', 108.
Pavement, accuracy of levelling of the Great Pyramid, 62.
Pavements, 62, 130.
Peg-dowels, 112, 151.
Penanhūret, statuette of, 65.
Pendant-leaf capital, 10.
Pent roofs, 12, 184, 189.
Petrie, Sir W. M. Flinders, 78, 94, 202, 214.
Philae, arch at, 188.
— pilaster and pavement at, 131.
— roof drainage at, 159.
— quarries for temples of, 15.
π (Pi), 118, 222.
Pick, mason's, 17, 31, 194, 202, 224.
Pigments, composition of ancient, 200.
Pilasters, 6, 9, 10, 131.
Pivots of doors, 163, 167.